Xuzhu Wang, Da Ruan and Etienne E. Kerre

Mathematics of Fuzziness – Basic Issues

Studies in Fuzziness and Soft Computing, Volume 245

Editor-in-Chief

Prof. Janusz Kacprzyk
Systems Research Institute
Polish Academy of Sciences
ul. Newelska 6
01-447 Warsaw
Poland
E-mail: kacprzyk@ibspan.waw.pl

Further volumes of this series can be found on our homepage: springer.com

Vol. 228. Adam Kasperski
Discrete Optimization with Interval Data,
2008
ISBN 978-3-540-78483-8

Vol. 230. Bhanu Prasad (Ed.)
Soft Computing Applications in Business,
2008
ISBN 978-3-540-79004-4

Vol. 231. Michal Baczynski,
Balasubramaniam Jayaram
Soft Fuzzy Implications, 2008
ISBN 978-3-540-69080-1

Vol. 232. Eduardo Massad,
Neli Regina Siqueira Ortega,
Laécio Carvalho de Barros,
Claudio José Struchiner
*Fuzzy Logic in Action: Applications
in Epidemiology and Beyond,* 2008
ISBN 978-3-540-69092-4

Vol. 233. Cengiz Kahraman (Ed.)
*Fuzzy Engineering Economics with
Applications,* 2008
ISBN 978-3-540-70809-4

Vol. 234. Eyal Kolman, Michael Margaliot
*Knowledge-Based Neurocomputing:
A Fuzzy Logic Approach,* 2009
ISBN 978-3-540-88076-9

Vol. 235. Kofi Kissi Dompere
Fuzzy Rationality, 2009
ISBN 978-3-540-88082-0

Vol. 236. Kofi Kissi Dompere
Epistemic Foundations of Fuzziness, 2009
ISBN 978-3-540-88084-4

Vol. 237. Kofi Kissi Dompere
Fuzziness and Approximate Reasoning, 2009
ISBN 978-3-540-88086-8

Vol. 238. Atanu Sengupta, Tapan Kumar Pal
*Fuzzy Preference Ordering of Interval
Numbers in Decision Problems,* 2009
ISBN 978-3-540-89914-3

Vol. 239. Baoding Liu
*Theory and Practice of Uncertain
Programming,* 2009
ISBN 978-3-540-89483-4

Vol. 240. Asli Celikyilmaz, I. Burhan Türksen
Modeling Uncertainty with Fuzzy Logic, 2009
ISBN 978-3-540-89923-5

Vol. 241. Jacek Kluska
*Analytical Methods in Fuzzy
Modeling and Control,* 2009
ISBN 978-3-540-89926-6

Vol. 242. Yaochu Jin, Lipo Wang
*Fuzzy Systems in Bioinformatics
and Computational Biology,* 2009
ISBN 978-3-540-89967-9

Vol. 243. Rudolf Seising (Ed.)
*Views on Fuzzy Sets and Systems from Different
Perspectives,* 2009
ISBN 978-3-540-93801-9

Vol. 244. Xiaodong Liu and Witold Pedrycz
*Axiomatic Fuzzy Set Theory and Its
Applications,* 2009
ISBN 978-3-642-00401-8

Vol. 245. Xuzhu Wang, Da Ruan,
Etienne E. Kerre
Mathematics of Fuzziness – Basic Issues, 2009
ISBN 978-3-540-78310-7

Xuzhu Wang, Da Ruan and Etienne E. Kerre

Mathematics of Fuzziness – Basic Issues

Springer

Authors

Prof. Dr. Xuzhu Wang
Department of Mathematics
Taiyuan University of Technology
Taiyuan, Shanxi
China
E-mail: wangxuzhu2006@126.com

Prof. Dr. Etienne E. Kerre
Department of Applied Mathematics &
Computer Science
Ghent University
Krijgslaan 281 (S9)
9000 Gent
Belgium
E-mail: etienne.kerre@ugent.be

Prof. Dr. Da Ruan
Belgian Nuclear Research Centre
(SCK • CEN)
Boeretang 200
2400 Mol
Belgium
E-mail: druan@sckcen.be

&

Department of Applied Mathematics &
Computer Science
Ghent University
Krijgslaan 281 (S9)
9000 Gent
Belgium
E-mail: da.ruan@ugent.be

ISBN 978-3-540-78310-7   e-ISBN 978-3-540-78311-4

DOI 10.1007/978-3-540-78311-4

Studies in Fuzziness and Soft Computing   ISSN 1434-9922

Library of Congress Control Number: 2009922105

© 2009 Springer-Verlag Berlin Heidelberg

This work is subject to copyright. All rights are reserved, whether the whole or part of the material is concerned, specifically the rights of translation, reprinting, reuse of illustrations, recitation, broadcasting, reproduction on microfilm or in any other way, and storage in data banks. Duplication of this publication or parts thereof is permitted only under the provisions of the German Copyright Law of September 9, 1965, in its current version, and permission for use must always be obtained from Springer. Violations are liable to prosecution under the German Copyright Law.

The use of general descriptive names, registered names, trademarks, etc. in this publication does not imply, even in the absence of a specific statement, that such names are exempt from the relevant protective laws and regulations and therefore free for general use.

Typeset & Cover Design: Scientific Publishing Services Pvt. Ltd., Chennai, India.

Printed in acid-free paper

9 8 7 6 5 4 3 2 1

springer.com

Foreword

The word *fuzziness,* as introduced by our friend and mentor Professor Lotfi A. Zadeh in 1965, has evolved to characterize the information associated with the human language. The human language which carries a lot of information and is being used in our everyday decision making processes, in a mathematical sense, is not very precise. For example, in Saskatoon in the month of January day time temperature is usually around -15^0 C to -20^0C. A resident of Saskatoon will classify this temperature as *'not very cold'*. By a friend just arriving from India, his classification for this temperature would be *'very- very cold'*.

Professor Zadeh has given us a mathematical tool called the *fuzzy mathematics,* which gives precision to uncertainty inherent in our human language. The meaning of the wordings that we use in our everyday life are context dependent, subjective, and have a cognitive flavor. Thus, we can say that *the fuzzy set theory is a mathematical tool that provides a figure of certainty to a cognitive type of uncertain phenomenon inherent in our human language.*

Since the inception of notion of fuzzy sets in 1965, many research papers and books have appeared in the field of fuzzy uncertainty. In the present book entitled *'Mathematics of Fuzziness, Basic Issues'* the authors, Drs. Xuzhu Wang, Da Ruan and Etienne Kerre who are among the *pioneers* in the field of fuzzy mathematics, have provided a lucid mathematical characterization of 'fuzzy uncertainty'. In this book the authors introduce a basic notion of *'fuzziness'* and provide a conceptual mathematical framework to characterize such fuzzy phenomena.

In general, mathematical methods have evolved in order to characterize the phenomena we are surrounded with. This book which contains six chapters in 281 pages lays the basic mathematical basis of fuzzy phenomena, the phenomena which are inherent in our human perception and cognitive processes. The authors start with the preliminaries on sets and relations, in Chapter 1. Then in Chapter 2 they provide some conceptual basics of fuzzy sets. In Chapter 3 they provide a detailed description of the methodology of fuzzy relations with applications in several fields such as fuzzy clustering, information retrieval and multiple attribute decision making analysis. Following this basic introduction, the authors move to Chapter 4 to describe the extension principle and fuzzy numbers, in Chapter 5 some mathematical topics such as fuzzy measures and fuzzy integrals, and finally in Chapter 6 some applications oriented topics such as fuzzy inference and fuzzy control. Each chapter is very well appended with exercises and reference material.

In the years ahead, the notion of *fuzziness* is likely to be an integral part for dealing with cognitive uncertainty in a computationally effective way. Thus, this book on mathematics of fuzziness which has evolved from the authors' lecture notes to both undergraduate and graduate students provides a basic mathematical exposition to this growing field of soft-computing. The authors and the publisher, Springer Verlag, have produced a treatise that addresses, with high authority and high level of expertise, the mathematics devoted in the studies of fuzziness and soft-computing. For this very informative and timely book the authors deserve our heartiest congratulations!

University of Saskatchewan   Madan M. Gupta
January 28, 2009

Preface

This book is intended to serve as a basic textbook for a fuzzy set and system course at the undergraduate level and as a preliminary material for further research and applications on some special topics. Accordingly, the readers may contain undergraduate as well as graduate students in mathematics, computer science and engineering, engineers and researchers in the related fields. The seven aspects feature our book.

(1) The most fundamental subjects in fuzzy set theory such as fuzzy sets, fuzzy relations and fuzzy numbers are introduced in detail.
(2) The most important applications of fuzzy sets such as pattern recognition, clustering analysis and fuzzy control are explicitly highlighted.
(3) The well-developed pure mathematical branches such as fuzzy measures, fuzzy integrals, fuzzy algebra, and fuzzy topology are briefly discussed.
(4) While most parts of the book are devoted to the introduction to the basics of fuzzy set theory and its applications, some parts contain advanced materials based our research.
(5) Remarks are made to indicate further reading in the end of some sections particularly with important and theoretical sections.
(6) The whole book is basically self-contained. There is no difficulty for the reader with basic knowledge of mathematical analysis and linear algebra to get through all chapters except Chapter 5.
(7) Many exercises arranged as the last section of every chapter help the reader to understand the concepts, approaches and theories in the chapter and serve as a supplement to main results in the chapter.

The book is structured as follows. In Chapter 1, we recall basic concepts of set theory and abstract algebra including set, relation, isomorphism, lattice, Boolean algebra, and soft algebra. These concepts play an important role in the investigation of fuzzy sets and fuzzy relations. In Chapter 2, we introduce the basics of fuzzy set theory. Considering general fuzzy logic connective operations are visible everywhere in the current fuzzy literature, a lot of space is assigned to deal with negations, t-norms, t-conorms, fuzzy implications and equivalencies besides the Zadeh's set-theoretic operations. Then we expose the links between fuzzy sets and crisp sets which include the decomposition theorems and mathematical representations of fuzzy sets in terms of a nest of sets. Afterwards, fuzzy sets are extended to L-fuzzy sets and a similar investigation is carried out. Finally, an important application of fuzzy set theory – fuzzy pattern recognition – is introduced. In Chapter 3, we deal with fuzzy relations. We begin with the investigation of various compositions of fuzzy relations and followed by the introduction of fuzzy equivalence relations and fuzzy tolerance relations. Then comes an exhaustive investigation into the main properties of fuzzy relations: negativity, semi-transitivity, Ferrers property, consistency, weak transitivity and acyclicity. After that, we discuss a type of composite fuzzy relation equations, and finally mention several applications of fuzzy relations. In Chapter 4, three tasks are fulfilled. The first one is the introduction of the Zadeh's extension principle. The second one is the discussion of fuzzy numbers and their algebraic operations. The last one is the detailed investigation of ranking methods for fuzzy numbers, which can find wide applications when fuzzy data are involved and processed. In Chapter 5, we introduce three well-developed fuzzified mathematical branches: (i) fuzzy measures and fuzzy integrals; (ii) fuzzy algebraic structures including fuzzy groups, fuzzy rings and fuzzy ideals; (iii) fuzzy topology. The chapter is especially ready for the reader with some crisp mathematical prerequisites and supplies them with primary fuzzification approaches of a mathematical theory, which will open a door to further theoretical research. The major subject of Chapter 6 is fuzzy control. We outline the principle of fuzzy control by a typical application case. Around fuzzy control, we briefly introduce some concepts related to fuzzy inference such as linguistic variables, hedges, fuzzy propositions, IF-THEN rules and special fuzzy inference approaches. Considering that fuzzification and defuzzification often play an important role in the design of fuzzy controllers, we also investigate them in this chapter.

This book has evolved from our lectures for both undergraduate and graduate students. We are thankful to our students and colleagues for their constructive suggestions. We are deeply indebted to the authors whose works are employed and not explicitly cited, but certainly listed in our bibliography. We gratefully acknowledge the research support by the China-Flanders Foundation and the Natural Science Foundation of Shanxi Province.

Xuzhu Wang
Da Ruan
Etienne E. Kerre

Contents

1 **Preliminaries** 1
   1.1 Sets 1
   1.2 Relations 3
   1.3 Mappings and Algebraic Systems 7
   1.4 Lattices 10
   1.5 Special Lattices 13
   1.6 Exercises 17

2 **Basics of Fuzzy Sets** 21
   2.1 Fuzzy Sets and Their Set-Theoretic Operations 21
   2.2 General Fuzzy Logic Connectives 26
      2.2.1 Fuzzy Negations 26
      2.2.2 Triangular Norms and Conorms 28
      2.2.3 Fuzzy Implications 32
      2.2.4 Fuzzy Equivalencies 38
   2.3 Decomposition of a Fuzzy Set 39
   2.4 Mathematical Representation of Fuzzy Sets 43
   2.5 L-Fuzzy Sets 46
      2.5.1 Pseudo-complements 46
      2.5.2 L-Fuzzy Sets and Their Set-Theoretic Operations 47
      2.5.3 Decomposition of an L-Fuzzy Set 49
      2.5.4 Mathematical Representation of L-Fuzzy Sets 51
   2.6 Fuzzy Pattern Recognition 54
      2.6.1 Type I Fuzzy Pattern Recognition 55
      2.6.2 Type II Fuzzy Pattern Recognition 56
   2.7 Exercises 60

3 Fuzzy Relations ... 65
3.1 Basic Concepts of Fuzzy Relations ... 65
3.2 Compositions of Fuzzy Relations ... 69
   3.2.1 Round Composition of Fuzzy Relations ... 69
   3.2.2 Subcomposition, Supercomposition and Square Composition of Fuzzy Relations ... 72
3.3 Fuzzy Equivalence Relations ... 74
3.4 Closures ... 77
   3.4.1 The Concept of a Closure ... 78
   3.4.2 The Transitive Closure of a Fuzzy Relation ... 79
3.5 Fuzzy Tolerance Relations ... 80
3.6 Other Special Fuzzy Relations ... 82
   3.6.1 Crisp Negative Transitivity, Semitransitivity, and Ferrers Property ... 83
   3.6.2 Negative S-Transitivity, T-S-Semitransitivity, and T-S-Ferrers Property ... 85
   3.6.3 Consistency, Weak Transitivity and Acyclicity ... 91
3.7 Fuzzy Relation Equations ... 94
3.8 Some Applications of Fuzzy Relations ... 99
   3.8.1 Fuzzy Clustering Analysis ... 99
   3.8.2 An Application to Information Retrieval ... 102
   3.8.3 An Application to Multiple Attribute Decision Making Analysis ... 104
3.9 Exercises ... 106

4 Extension Principle and Fuzzy Numbers ... 113
4.1 Unary Extension Principle ... 114
4.2 n-Ary Extension Principle ... 119
4.3 Convex Fuzzy Quantities ... 122
4.4 Fuzzy Numbers ... 126
   4.4.1 The Concept of a Fuzzy Number ... 126
   4.4.2 Properties of Algebraic Operations on Fuzzy Numbers ... 130
4.5 Ranking of Fuzzy Numbers ... 132
   4.5.1 Ranking Fuzzy Numbers by a Ranking Function ... 133
   4.5.2 Ranking Fuzzy Numbers According to the Closeness to a Reference Set ... 134
   4.5.3 Ranking Fuzzy Numbers Based on Pairwise Comparisons ... 136
   4.5.4 Ranking Axioms ... 140
4.6 An Application of Fuzzy Numbers ... 142
4.7 Exercises ... 146

Contents

5 A Brief Introduction to Some Pure Mathematical Topics 153
   5.1 Fuzzy Measures and Fuzzy Integrals 153
       5.1.1 Fuzzy Measures 153
       5.1.2 Fuzzy Integrals 157
   5.2 Fuzzy Algebra 163
       5.2.1 Fuzzy Subgroups 163
       5.2.2 Normal Fuzzy Subgroups 167
       5.2.3 Fuzzy Subrings 171
       5.2.4 Fuzzy Ideals 173
   5.3 Fuzzy Topology 175
       5.3.1 Definitions 175
       5.3.2 Characterization of a Fuzzy Topology in Terms of Preassigned Operations 177
       5.3.3 Characterization of a Fuzzy Topology in Terms of Closed Sets 179
       5.3.4 Characterization of a Fuzzy Topology Using the Interior Operator 180
       5.3.5 Characterization of a Fuzzy Topology by Means of a Closure Operator 182
       5.3.6 Characterization of a Fuzzy Topology by Means of Neighborhood Systems 182
       5.3.7 Normality in Fuzzy Topological Spaces 184
       5.3.8 Some Examples of Fuzzy Topological Spaces 185
   5.4 Exercises 186

6 Fuzzy Inference and Fuzzy Control 189
   6.1 Linguistic Variables and Hedges 189
   6.2 Fuzzy Propositions and IF-THEN Rules 191
   6.3 Fuzzy Inference Rules 193
   6.4 The Calculation of Inference Results 194
   6.5 Fuzzification and Defuzzification 198
   6.6 The Principle of Fuzzy Control 201
   6.7 Exercises 204

References 207

Index 217

Chapter 1
Preliminaries

This chapter aims to recall basic concepts of set theory and abstract algebra including set, relation, isomorphism, lattice, Boolean algebra and soft algebra, which will serve as the base of the remaining chapters in the book.

1.1 Sets

A set is viewed as a collection of objects satisfying certain desired properties. Every object in the set is mathematically called an element or member. If every element of set A is also an element of set B, then A is called a subset of B and this is written as $A \subseteq B$. If $A \subseteq B$ and $B \subseteq A$, then we say that A and B are equal, written as $A = B$. If $A \subseteq B, A \neq B$ and $A \neq \emptyset$, where $\emptyset$ denotes the empty set, then A is called a proper subset of B, denoted by $A \subset B$. Let X be the universe of discourse (or the universe for short). The set of all subsets of X is denoted by $P(X)$, i.e. $P(X) = \{A | A \subseteq X\}$, and is called the power set (class) of X. For $A, B \in P(X)$, some of the set-theoretic operations are defined as follows:

The union of A and B is defined by

$$A \cup B = \{x | x \in A \text{ or } x \in B\}.$$

The intersection of A and B is defined by

$$A \cap B = \{x | x \in A \text{ and } x \in B\}.$$

The complement of A is defined by

$$A^c = \{x | x \in X \text{ and } x \notin A\}.$$

Proposition 1.1. *The above defined set-theoretic operations satisfy* ($\forall A, B, C \in P(X)$):

(1) idempotency: $A \cup A = A$, $A \cap A = A$;
(2) commutativity: $A \cup B = B \cup A$, $A \cap B = B \cap A$;

(3) *associativity:* $(A \cup B) \cup C = A \cup (B \cup C)$, $(A \cap B) \cap C = A \cap (B \cap C)$;
(4) *absorption laws:* $A \cup (A \cap B) = A$, $A \cap (A \cup B) = A$;
(5) *distributivity:* $A \cup (B \cap C) = (A \cup B) \cap (A \cup C)$, $A \cap (B \cup C) = (A \cap B) \cup (A \cap C)$;
(6) *the existence of the greatest and least element:* $\emptyset \subseteq A \subseteq X$.
(7) *involution:* $(A^c)^c = A$;
(8) *De Morgan laws:* $(A \cup B)^c = A^c \cap B^c$, $(A \cap B)^c = A^c \cup B^c$;
(9) *complementation:* $A \cup A^c = X$ *(the law of excluded middle),*
$$A \cap A^c = \emptyset \text{ (the law of contradiction)}.$$

Proof. We only prove the first equality of (8) as an illustration. The remaining proofs are left to the reader.

$\forall x \in X$, $x \in (A \cup B)^c \Leftrightarrow x \notin A \cup B \Leftrightarrow (x \notin A \text{ and } x \notin B) \Leftrightarrow x \in A^c \cap B^c$.
More generally, the union and intersection of A_i ($i \in I$) may be similarly defined with an arbitrary index set I.

$$\bigcup_{i \in I} A_i = \{x | \exists i \in I, x \in A_i\};$$

$$\bigcap_{i \in I} A_i = \{x | \forall i \in I, x \in A_i\}.$$

Accordingly, the associativity, distributivity and De Morgan laws are extended as

$$(\bigcup_{i \in I} A_i) \cup (\bigcup_{i \in I} B_i) = \bigcup_{i \in I}(A_i \cup B_i) \text{ and } (\bigcap_{i \in I} A_i) \cap (\bigcap_{i \in I} B_i) = \bigcap_{i \in I}(A_i \cap B_i),$$

$$A \cup (\bigcap_{i \in I} A_i) = \bigcap_{i \in I}(A \cup B_i) \text{ and } A \cap (\bigcup_{i \in I} B_i) = \bigcup_{i \in I}(A \cap B_i),$$

$$(\bigcup_{i \in I} A_i)^c = \bigcap_{i \in I}(A_i)^c \text{ and } (\bigcap_{i \in I} A_i)^c = \bigcup_{i \in I}(A_i)^c,$$

where I is an arbitrary index set and $A, A_i, B_i \in P(X)(\forall i \in I)$.

In describing sets, an important tool is the characteristic function. Let A be a subset of X. The characteristic function of A is defined by: $\forall x \in X$,

$$\chi_A(x) = \begin{cases} 1 & x \in A; \\ 0 & \text{otherwise}. \end{cases}$$

The set-theoretic operations of union, intersection and complement may be expressed by means of characteristic functions.

Proposition 1.2. *The characteristic functions of $A \cup B$, $A \cap B$ and A^c are calculated by using the following equalities for $x \in X$:*

(1) $\chi_{A \cup B}(x) = \max(\chi_A(x), \chi_B(x))$;
(2) $\chi_{A \cap B}(x) = \min(\chi_A(x), \chi_B(x))$;
(3) $\chi_{A^c}(x) = 1 - \chi_A(x)$.

Proof. (1) $\chi_{A\cup B}(x) = 1 \Leftrightarrow x \in A \cup B \Leftrightarrow x \in A$ or $x \in B \Leftrightarrow \chi_A(x) = 1$ or $\chi_B(x) = 1 \Leftrightarrow \max(\chi_A(x), \chi_B(x)) = 1$. Hence,

$$\chi_{A\cup B}(x) = \max(\chi_A(x), \chi_B(x)).$$

The proof of (2) and (3) is similar. □

Remark 1.1. *The first two equalities in Proposition 1.2 can be extended as*

$$\forall x, \chi_{\cup_{i\in I} A_i}(x) = \sup_{i\in I} \chi_{A_i}(x),$$

$$\forall x, \chi_{\cap_{i\in I} A_i}(x) = \inf_{i\in I} \chi_{A_i}(x),$$

where I is an arbitrary index set. We leave the proofs to the reader.

In addition, it is easily checked that

$$A \subseteq B \Leftrightarrow \forall x, \chi_A(x) \leq \chi_B(x)$$

$$A = B \Leftrightarrow \forall x, \chi_A(x) = \chi_B(x).$$

1.2 Relations

Let X and Y be two sets. A subset R of $X \times Y$, i.e. $R \in P(X \times Y)$, is called a relation from X to Y. If $(x, y) \in R$, we simply write xRy. By definition, a relation R from X to Y is a set. If the set is empty, then R is called the empty relation. Another special relation from X to Y is $X \times Y$, which is called the entire relation.

Example 1.1. *Let $X = \{1, 2, 3, 4\}$, $Y = \{2, 3, 4, 5\}$. The relation R_1 "less than" from X to Y is defined as*

$$R_1 = \{(1,2), (1,3), (1,4), (1,5), (2,3), (2,4), (2,5), (3,4), (3,5), (4,5)\},$$

while the relation R_2 "less than" from Y to X is defined as

$$R_2 = \{(2,3), (2,4), (3,4)\}.$$

For a relation R from X to Y, $x \in X$ and $y \in Y$, we define the R-afterset of x, denoted by xR, as $xR = \{y | y \in Y, (x, y) \in R\}$ and the R-foreset of y, denoted by Ry, as $Ry = \{x | x \in X, (x, y) \in R\}$. Clearly, the R-afterset (R-foreset) of x is a subset of Y (resp. X). For instance, in Example 1.1, $1R_1 = \{2, 3, 4, 5\}$, $2R_1 = \{3, 4, 5\}$, $R_1 2 = \{1\}$, and $R_1 3 = \{1, 2\}$.

A relation from X to X is said to be a (binary) relation on X. For instance, let $X = \{1, 2, 3, 4, 5, 6\}$. Define the relation R by: xRy iff x divides y (or $x|y$ in symbols). Then $R = \{(1,1), (1,2), (1,3), (1,4), (1,5), (1,6), (2,2), (3,3),$ $(4,4), (5,5), (6,6), (2,4), (2,6), (3,6)\}$ is a relation on X. A special relation on X is $I_X = \{(x,x)|x \in X\}$, which is referred to as the identity relation. Since relations are sets, many notions of relations are naturally similar to those of sets, among which we mention the following (R, R_1 and R_2 are relations from X to Y):

(1) $R_1 \subseteq R_2$ iff xR_1y implies xR_2y for all $(x,y) \in X \times Y$;
(2) $R_1 = R_2$ iff $R_1 \subseteq R_2$ and $R_2 \subseteq R_1$;
(3) The union of R_1 and R_2 is the relation

$$R_1 \cup R_2 = \{(x,y)|xR_1y \text{ or } xR_2y\};$$

(4) The intersection of R_1 and R_2 is the relation

$$R_1 \cap R_2 = \{(x,y)|xR_1y \text{ and } xR_2y\}.$$

(5) The complement of R is the relation $R^c = \{(x,y)|(x,y) \notin R\}$.

The properties of relation operations are the same as those of set operations. As a result, all the equalities in Proposition 1.1 are true for relations. Besides the union, intersection and complement, relations have their own operations. Among them are the inverse and composition. Let R be a relation from X to Y. The inverse R^{-1} of R is a relation from Y to X defined by

$$(y,x) \in R^{-1} \text{ iff } (x,y) \in R.$$

For two relations R_1 from X to Y and R_2 from Y to Z, the composition $R_1 \circ R_2$ (read as: R_1 followed by R_2 or R_1 before R_2) of R_1 and R_2 is the relation from X to Z defined by

$$R_1 \circ R_2 = \{(x,z)|\exists y \in Y, xR_1y \text{ and } yR_2z\},$$

or equivalently, $R_1 \circ R_2 = \{(x,z)|xR_1 \cap R_2z \neq \emptyset\}$ in terms of afterset and foreset. Specifically, $R \circ R$ is simply written as R^2 when R is a relation on X. More generally, R^n is recursively defined as $R^{n-1} \circ R$ for every positive integer n. It is easily checked that

(1) $(R^{-1})^c = (R^c)^{-1}$;
(2) $(R_1 \cup R_2)^{-1} = R_1^{-1} \cup R_2^{-1}$, $(R_1 \cap R_2)^{-1} = R_1^{-1} \cap R_2^{-1}$;
(3) $(R_1 \circ R_2) \circ R_3 = R_1 \circ (R_2 \circ R_3)$;
(4) $(R_1 \circ R_2)^{-1} = R_2^{-1} \circ R_1^{-1}$;
(5) $R^l \circ R^m = R^{l+m}$, $(R^l)^m = R^{lm}$, where l, m, n are arbitrary positive integers.

Remark 1.2. *If one defines $R^{-n} = (R^{-1})^n$ for any positive integer n, then (4) is valid for any integers l, m, n.*

Based on the notation, the composition is sometimes called the round composition (product). There also exist other types of compositions like triangular compositions (products) and square composition (product), which are defined

1.2 Relations

as follows. Assume that R_1 is a relation from X to Y and R_2 is a relation from Y to Z.

The subcomposition (subproduct) of R_1 and R_2 is a relation from X to Z defined by

$$R_1 \triangleleft R_2 = \{(x,z) | xR_1 \cap R_2 z \neq \emptyset \text{ and } xR_1 \subseteq R_2 z\}.$$

The supercomposition (superproduct) of R_1 and R_2 is a relation from X to Z defined by

$$R_1 \triangleright R_2 = \{(x,z) | xR_1 \cap R_2 z \neq \emptyset \text{ and } R_2 z \subseteq xR_1\}.$$

The subcomposition and supercomposition are called triangular compositions.

The square composition of R_1 and R_2 is a relation from X to Z defined by

$$R_1 \square R_2 = \{(x,z) | xR_1 \cap R_2 z \neq \emptyset \text{ and } xR_1 = R_2 z\}.$$

From the definitions, it follows immediately that

(1) $R_1 \square R_2 = (R_1 \triangleleft R_2) \cap (R_1 \triangleright R_2)$ and $R_1 \square R_2 = R_1^c \square R_2^c$;
(2) $R_1 \triangleleft R_2 = (R_2^{-1} \triangleright R_1^{-1})^{-1}$ and $R_1 \triangleleft R_2 = R_1^c \triangleright R_2^c$;
(3) $R_1 \triangleright R_2 = (R_2^{-1} \triangleleft R_1^{-1})^{-1}$ and $R_1 \triangleright R_2 = R_1^c \triangleleft R_2^c$.

Clearly, the characteristic function χ_R of a relation R from X to Y can be explicitly written as

$$\chi_R(x,y) = \begin{cases} 1 & xRy; \\ 0 & \text{otherwise}. \end{cases}$$

If $R, S \in P(X \times Y)$, $T \in P(Y \times Z)$, then the characteristic functions of some operations are listed in the following:

(1) $\forall (x,y) \in X \times Y, \chi_{R \cup S}(x,y) = \max(\chi_R(x,y), \chi_S(x,y))$;
(2) $\forall (x,y) \in X \times Y, \chi_{R \cap S}(x,y) = \min(\chi_R(x,y), \chi_S(x,y))$;
(3) $\forall (x,y) \in X \times Y, \chi_{R^c}(x,y) = 1 - \chi_R(x,y)$;
(4) $\forall (x,y) \in X \times Y, \chi_{R^{-1}}(y,x) = \chi_R(x,y)$;
(5) $\forall (x,z) \in X \times Z, \chi_{R \circ T}(x,z) = \sup_{y \in Y} \min(\chi_R(x,y), \chi_T(y,z))$.

Proof. The first three equalities are direct consequences of the expression of the characteristic function of the set-theoretic operations. The proof of (4) is trivial. Hence, we only verify the equality (5).

Indeed, $\forall (x,z) \in X \times Z$,

$$\chi_{R \circ T}(x,z) = 1 \Leftrightarrow (x,z) \in R \circ T$$
$$\Leftrightarrow \exists y \in Y, (x,y) \in R \text{ and } (y,z) \in T$$
$$\Leftrightarrow \exists y \in Y, \chi_R(x,y) = 1 \text{ and } \chi_T(y,z) = 1$$
$$\Leftrightarrow \exists y \in Y, \min(\chi_R(x,y), \chi_T(y,z)) = 1$$
$$\Leftrightarrow \sup_{y \in Y} \min(\chi_R(x,y), \chi_T(y,z)) = 1.$$

These equivalencies indicate that

$$\forall (x,z) \in X \times Z, \chi_{R \circ T}(x,z) = \sup_{y \in Y} \min(\chi_R(x,y), \chi_T(y,z)). \qquad \square$$

Of all relations, the most important is the so-called equivalence relation, which is useful particularly in clustering analysis.

Definition 1.1. *A relation R on X is called an equivalence relation if it satisfies:*

(1) reflexivity: $\forall x \in X$, xRx;
(2) symmetry: $\forall x, y \in X$, xRy implies yRx;
(3) transitivity: $\forall x, y, z \in X$, xRy and yRz imply xRz.

Remark 1.3. *Clearly, R is symmetric iff $R = R^{-1}$, and R is transitive iff $R^2 \subseteq R$.*

Example 1.2. *Let $X = \{a, b, c, d, e\}$. The relation R on X is defined as $R = \{(a,a), (b,b), (c,c), (d,d), (e,e), (a,b), (a,c), (b,a), (b,c), (c,a), (c,b), (d,e), (e,d)\}$. Then it can be checked that R is an equivalence relation.*

Let R be an equivalence relation on X. Write $[x]_R = \{y | yRx\}$, which is simply written as $[x]$ if no confusion is possible. Then $[x]$ is called the equivalence class with x as a representative element. Note that $[x]_R = xR = Rx$. The set of all equivalence classes is called the quotient set of X by R, which is denoted by X/R, namely $X/R = \{[x] | x \in X\}$. For instance, in Example 1.2, $[a] = \{a, b, c\} = [b] = [c]$ and $[d] = \{d, e\} = [e]$. Hence, the quotient set of X by R is $X/R = \{[a], [d]\}$. In this example, the union of all equivalence classes is X, and different equivalence classes are disjoint. These properties hold in general.

Proposition 1.3. *Let R be an equivalence relation on X. Then*

(1) $\bigcup_{x \in X} [x] = X$;
(2) $\forall x_1, x_2 \in X$, $[x_1] = [x_2]$ or $[x_1] \cap [x_2] = \emptyset$.

Proof. (1) is trivial. We prove (2). Suppose that $[x_1] \cap [x_2] \neq \emptyset$. Then $\exists x \in [x_1] \cap [x_2]$. Hence $x_1 R x$ and $x_2 R x$. By the symmetry and transitivity of R, we have $x_1 R x_2$. Now we show that $[x_1] \subseteq [x_2]$. Indeed, $x \in [x_1]$ implies $x_1 R x$, which, together with $x_1 R x_2$, indicates xRx_2 by the transitivity of R. Therefore, $[x_1] \subseteq [x_2]$. Similarly, $[x_2] \subseteq [x_1]$, and thus $[x_1] = [x_2]$. $\qquad \square$

Given an equivalence relation on X, Proposition 1.3 indicates that the set of all equivalence classes constitutes a partition of X.

Finally, we focus on relations on finite universes of discourse. More specifically, let $X = \{x_1, x_2, \ldots, x_m\}$, $Y = \{y_1, y_2, \ldots, y_n\}$, and R a relation from X to Y. In this situation, write $r_{ij} = \chi_R(x_i, y_j)$ ($i = 1, 2, \ldots, m; j = 1, 2, \ldots, n$). Then R is uniquely determined by the matrix $(r_{ij})_{m \times n}$. We shall not distinguish R from this matrix. In other words, we simply write $R = (r_{ij})_{m \times n}$, where $r_{ij} = 1$ or 0 according as $x_i R y_j$ or $x_i R^c y_j$. For example, the relation R in Example 1.2 can be represented by

$$\begin{pmatrix} 1 & 1 & 1 & 0 & 0 \\ 1 & 1 & 1 & 0 & 0 \\ 1 & 1 & 1 & 0 & 0 \\ 0 & 0 & 0 & 1 & 1 \\ 0 & 0 & 0 & 1 & 1 \end{pmatrix}$$

With the matrix representation of a relation, set-theoretic operations of relations can be readily performed. Let $R = (r_{ij})_{m \times n}$ and $S = (s_{ij})_{m \times n}$. Then $R \cup S = (t_{ij})_{m \times n}$, where $t_{ij} = \max(r_{ij}, s_{ij})$. Similarly, $R \cap S = (\min(r_{ij}, s_{ij}))_{m \times n}$, $R^c = (1 - r_{ij})_{m \times n}$, $R^{-1} = (r'_{ij})_{n \times m}$, where $r'_{ij} = r_{ji}$. Hence, R^{-1} is the transpose of R in terms of matrix. In addition, from the characteristic function expression of (round) composition, it follows that $R \circ S = (t_{ij})_{l \times n}$, where

$$t_{ij} = \max\{\min(r_{i1}, s_{1j}), \min(r_{i2}, s_{2j}), \cdots, \min(r_{im}, s_{mj})\}$$

(written simply as $\bigvee_{k=1}^{m}(r_{ik} \wedge s_{kj}))$ if $R = (r_{ik})_{l \times m}$ and $S = (s_{kj})_{m \times n}$. The computation of $R \triangleleft S$, $R \triangleright S$ and $R \square S$ is left as an exercise (Exercise 11).

1.3 Mappings and Algebraic Systems

Let X and Y be two sets. To each x in X, by some rule f, corresponds one and only one definite element in Y. Then f is called a mapping from X to Y. In symbols, $f : X \to Y$, $\forall x \in X$, $x \mapsto f(x)$. For example, the characteristic function χ_A ($A \in P(X)$) is a mapping from X to $\{0, 1\}$; the cardinality of A defines a mapping from $P(X)$ to $\mathbb{Z}^+$ (the set of all non-negative integers) as A changes in $P(X)$; each function of a single variable in mathematical analysis is a mapping from $\mathbb{R}$ (the set of all real numbers) to $\mathbb{R}$.

Clearly, if f is a mapping from X to Y and $Y \subseteq Z$, then f can be also regarded as a mapping from X to Z. For instance, χ_A is a mapping from X to $[0, 1]$ as well.

Let f be a mapping from X to Y. If $f(x) = y$, we say that y is the image under f of x. The image $f(A)$ under f of a set $A \in P(X)$ is the set of the images of all elements in A, i.e.

$$f(A) = \{y | \exists x \in A, y = f(x)\}.$$

If f_1 is a mapping from X to Y and f_2 is a mapping from Y to Z, then their composition $f_2 \circ f_1$ is the mapping from X to Z defined by:

$$\forall x \in X, (f_2 \circ f_1)(x) = f_2(f_1(x)).$$

It follows immediately that $f_3 \circ (f_2 \circ f_1) = (f_3 \circ f_2) \circ f_1$, i.e. the composition is associative.

Definition 1.2. *Let f be a mapping from X to Y. If $x_1 \neq x_2$ implies $f(x_1) \neq f(x_2)$, then f is called injective (or an injection). If $\forall y \in Y$, there exists an $x \in X$ such that $f(x) = y$, then f is called surjective (or a surjection). If f is injective and surjective, then f is called bijective (or a bijection).*

Clearly, for a mapping f from X to Y, (1) f is injective iff $\forall x_1, x_2 \in X$, $f(x_1) = f(x_2)$ implies $x_1 = x_2$; (2) f is surjective iff $f(X) = Y$.

Definition 1.3. *If $S \neq \emptyset$, a mapping $f : \overbrace{S \times S \times \cdots \times S}^{n} \to S$ is called an n-ary operation on S. Particularly, f is called an algebraic operation on S when $n = 2$.*

For example, $+, -, \times, \max, \min$ are all algebraic operations on $\mathbb{R}$. The union and intersection are algebraic operations on $P(X)$ and the complementation is a unary operation on $P(X)$.

Definition 1.4. *Let $f_1, f_2, \cdots f_m$ be m operations on a non-empty set S. Then $(S, f_1, f_2, \cdots f_m)$ is called an algebraic system.*

$(R, +, -, \times, \max, \min)$ and $(P(X), \cup, \cap, c)$ are examples of algebraic systems.

Example 1.3. Let $Ch(X) = \{\chi | \chi : X \to \{0, 1\}\}$. The operations $\vee, \wedge, c$ are defined as follows:

$\vee : Ch(X) \times Ch(X) \to Ch(X)$
$(\chi_1, \chi_2) \mapsto \chi_1 \vee \chi_2$
$\forall x \in X, (\chi_1 \vee \chi_2)(x) = \max(\chi_1(x), \chi_2(x));$
$\wedge : Ch(X) \times Ch(X) \to Ch(X)$
$(\chi_1, \chi_2) \mapsto \chi_1 \wedge \chi_2$
$\forall x \in X, (\chi_1 \wedge \chi_2)(x) = \min(\chi_1(x), \chi_2(x));$
$c : Ch(X) \to Ch(X)$
$\chi \mapsto c(\chi)$ (written as χ^c usually)
$\forall x \in X, \chi^c(x) = 1 - \chi(x).$

Then $(Ch(X), \vee, \wedge, c)$ is an algebraic system.

Definition 1.5. *Let $(A, f_1, f_2, \cdots, f_m)$ and $(B, g_1, g_2, \cdots, g_m)$ be two algebraic systems, where both f_i and g_i are n_i-ary operations $(i = 1, 2, \cdots, m)$. If there exists a mapping $f : A \to B$ such that $\forall a_1, a_2, \cdots, a_{n_i} \in A$, $f(f_i(a_1, a_2, \cdots, a_{n_i})) = g_i(f(a_1), f(a_2), \cdots, f(a_{n_i}))$, then the algebraic systems $(A, f_1, f_2, \cdots, f_m)$ and $(B, g_1, g_2, \cdots, g_m)$ are called homomorphic and f is called a homomorphism.*

Remark 1.4. *A homomorphism is frequently called an operation-preserving mapping in view that it preserves all operations.*

1.3 Mappings and Algebraic Systems

Definition 1.6. *Let $(A, f_1, f_2, \cdots, f_m)$ and $(B, g_1, g_2, \cdots, g_m)$ be two homomorphic algebraic systems, and f be the corresponding homomorphism. If f is surjective, then f is called an epimorphism. If, furthermore, f is bijective, then the two algebraic systems are called isomorphic and f is called an isomorphism.*

Example 1.4. *The algebraic systems $(P(X), \cup, \cap, c)$ and $(Ch(X), \vee, \wedge, c)$ are isomorphic.*

Let $f: P(X) \to Ch(X)$ be defined by $\forall A \in P(X)$, $f(A) = \chi_A$. We prove that f is an isomorphism between the two algebraic systems under consideration.

If $f(A) = f(B)$, then $\chi_A = \chi_B$, and thus $A = B$. Hence f is an injection.

For every $\chi \in Ch(X)$, write $A = \{x | \chi(x) = 1\}$. Then $f(A) = \chi_A = \chi$ (Indeed, $\chi_A(x) = 1 \Leftrightarrow x \in A \Leftrightarrow \chi(x) = 1$). Consequently, f is surjective.

Since $\chi_{A \cup B}(x) = \max\{\chi_A(x), \chi_B(x)\} = (\chi_A \vee \chi_B)(x)$, $f(A \cup B) = \chi_{A \cup B} = \chi_A \vee \chi_B = f(A) \vee f(B)$. Similarly

$$f(A \cap B) = f(A) \wedge f(B) \text{ and } f(A^c) = (f(A))^c.$$

Therefore f is an isomorphism. □

Since two isomorphic algebraic systems are mathematically viewed as the same, a subset of X can be understood as a mapping from X to $\{0, 1\}$.

Let $(A, f_1, f_2, \cdots, f_m)$ and $(B, g_1, g_2, \cdots, g_m)$ be two homomorphic algebraic systems with the homomorphism f. Suppose that both f_i and g_i ($i = 1, 2, \cdots, m$) are n_i-ary operations. Define the relation R_f on A by: $\forall a_1, a_2 \in A$, $a_1 R_f a_2 \iff f(a_1) = f(a_2)$. It is easily checked that R_f is an equivalence relation. Furthermore, define the n_i-ary operation f_i' on A/R_f by:

$$\forall [a_1], [a_2], \cdots, [a_{n_i}] \in A/R_f, f_i'([a_1], [a_2], \cdots, [a_{n_i}]) = [f_i(a_1, a_2, \cdots, a_{n_i})].$$

Concerning f_i', we have the following conclusion.

Proposition 1.4. *If f is a homomorphism, then f_i' is well-defined.*

Proof. It suffices to prove that the definition of f_i is independent of the choice of representative element. Indeed, if $[a_j] = [a_j']$ ($j = 1, 2, \cdots, n_i$), then $f(a_j) = f(a_j')$. Since f is a homomorphism,

$$\begin{aligned} f(f_i(a_1, a_2, \cdots, a_{n_i})) &= g_i(f(a_1), f(a_2), \cdots, f(a_{n_i})) \\ &= g_i(f(a_1'), f(a_2'), \cdots, f(a_{n_i}')) \\ &= f(f_i(a_1', a_2', \cdots, a_{n_i}')). \end{aligned}$$

Hence $[f_i(a_1, a_2, \cdots, a_{n_i})] = [f_i(a_1', a_2', \cdots, a_{n_i}')]$. □

Now comes the new algebraic system $(A/R_f, f_1', f_2', \cdots, f_m')$. Define the mapping f' from A/R_f to B by: $\forall [a] \in A/R_f$, $f'([a]) = f(a)$. Then we have:

Proposition 1.5. *If f is a surjective homomorphism between the algebraic systems $(A, f_1, f_2, \cdots, f_m)$ and $(B, g_1, g_2, \cdots, g_m)$, then f' is an isomorphism between the algebraic systems $(A/R_f, f'_1, f'_2, \cdots, f'_m)$ and $(B, g_1, g_2, \cdots, g_m)$.*

Proof. Firstly, if $f'([a_1]) = f'([a_2])$, then $f(a_1) = f(a_2)$. By the definition of R_f, $a_1 R_f a_2$, and thus $[a_1] = [a_2]$. Therefore f' is injective.

Secondly, since f is surjective, $\forall b \in B$, $\exists a \in A$ such that $f(a) = b$. As a result, $f'([a]) = f(a) = b$, and hence f' is surjective.

Finally, since f is a homomorphism, we have successively,

$$\begin{aligned} f'(f'_i([a_1],[a_2],\cdots,[a_{n_i}])) &= f'([f_i(a_1,a_2,\cdots,a_{n_i})]) \\ &= f(f_i(a_1,a_2,\cdots,a_{n_i})) \\ &= g_i(f(a_1),f(a_2),\cdots,f(a_{n_i})) \\ &= g_i(f'([a_1]),f'([a_2]),\cdots,f'([a_{n_i}])), \end{aligned}$$

i.e. f' is a homomorphism. Consequently, f' is an isomorphism, which completes the proof of the proposition. $\square$

1.4 Lattices

Definition 1.7. *Let $\leq$ be a binary relation on a non-empty set P. $\leq$ is called a partial order if $\leq$ satisfies:*

(1) reflexivity: $(\forall \alpha \in P)$ $(\alpha \leq \alpha)$;
(2) antisymmetry: $(\forall (\alpha, \beta) \in P^2)$ $(\alpha \leq \beta$ and $\beta \leq \alpha \Rightarrow \alpha = \beta)$;
(3) transitivity: $(\forall (\alpha, \beta, \gamma) \in P^3)$ $(\alpha \leq \beta$ and $\beta \leq \gamma \Rightarrow \alpha \leq \gamma)$.

An ordered pair $(P, \leq)$ consisting of a set and a partial order on the set is called a partially ordered set or a poset for short.

Two simple examples of posets are $(P(X), \subseteq)$ and $(\mathbb{R}, \leq)$, where $\leq$ is the "less than or equal to" between real numbers. In a poset, for convenience, $\alpha \leq \beta$ is alternatively written as $\beta \geq \alpha$. If $\alpha \leq \beta$ and $\alpha \neq \beta$, then we simply write $\alpha < \beta$ (or $\beta > \alpha$).

Example 1.5. *Let $\mathbb{Z}^+$ be the set of all non-negative integers. Define " $\leq$ " by: $a \leq b$ iff $a|b$. It is easily verified that $(\mathbb{Z}^+, \leq)$ is a poset. We will denote this poset by $(\mathbb{Z}^+, |)$.*

In Example 1.5, neither $2 \leq 3$ nor $3 \leq 2$ holds. If a partially ordered set $(P, \leq)$ satisfies: $\forall a, b \in P$, $a \leq b$ or $b \leq a$, then $(P, \leq)$ is called a totally ordered set. For instance, $(\mathbb{R}, \leq)$ is a totally ordered set while $(P(X), \subseteq)$ is not.

Definition 1.8. *Let $(P, \leq)$ be a poset and $S \subseteq P$.*

(1) If there exists an $a \in P$ such that $\forall x \in S$, $x \leq a$, then a is called an upper bound of S. If a is an upper bound of S and if $a \leq b$ holds for any other upper bound b of S, then a is called the least upper bound or supremum of S.

1.4 Lattices

(2) If there exists an $a \in P$ such that $\forall x \in S$, $a \leq x$, then a is called a lower bound of S. If a is a lower bound of S and if $b \leq a$ holds for any other lower bound b of S, then a is called the greatest lower bound or infimum of S.

Clearly, if the supremum of S exists, then it is unique, which will be denoted by $\sup S$ for brevity. Similarly, the infimum of S is denoted by $\inf S$ if it does exist.

The supremum and infimum are generalizations of the corresponding notions in $\mathbb{R}$. As we know, there always exist the supremum and infimum for a bounded non-empty set of real numbers. Furthermore, if S is a finite non-empty set of real numbers, then the supremum of S is the greatest one of all numbers in S and the infimum is the smallest one. However, it is more complicated in the general case.

Example 1.6. *Consider $S = \{2, 3\}$ in Example 1.5. There are infinitely many upper bounds of S including 6,12,18 etc. and the least upper bound is 6, i.e. $\sup S = 6$. However, 1 is the only one lower bound of S, and thus $\inf S = 1$. More generally, if $S = \{a, b\}$, then it can be verified that the supremum of S is the least common multiple of a and b, and the infimum of S is the the greatest common factor of a and b.*

Example 1.6 indicates that the supremum or infimum of S is not necessarily one of elements in S even though S is finite.

Proposition 1.6. *If $(P, \leq)$ is a poset, then:*

(1) $\sup\{\sup\{\alpha, \beta\}, \gamma\} = \sup\{\alpha, \sup\{\beta, \gamma\}\} = \sup\{\alpha, \beta, \gamma\}$;
(2) $\inf\{\inf\{\alpha, \beta\}, \gamma\} = \inf\{\alpha, \inf\{\beta, \gamma\}\} = \inf\{\alpha, \beta, \gamma\}$;
(3) $\alpha \leq \beta \Leftrightarrow \sup\{\alpha, \beta\} = \beta \Leftrightarrow \inf\{\alpha, \beta\} = \alpha$.

Proof. (1) From $\alpha \leq \sup\{\alpha, \beta\}$ and $\beta \leq \sup\{\alpha, \beta\}$, it follows that

$$\alpha \leq \sup\{\sup\{\alpha, \beta\}, \gamma\} \quad \text{and} \quad \beta \leq \sup\{\sup\{\alpha, \beta\}, \gamma\}.$$

In addition, $\gamma \leq \sup\{\sup\{\alpha, \beta\}, \gamma\}$. Hence

$$\sup\{\alpha, \beta, \gamma\} \leq \sup\{\sup\{\alpha, \beta\}, \gamma\}.$$

On the other hand, $\alpha \leq \sup\{\alpha, \beta, \gamma\}$ and $\beta \leq \sup\{\alpha, \beta, \gamma\}$ imply $\sup\{\alpha, \beta\} \leq \sup\{\alpha, \beta, \gamma\}$. Together with $\gamma \leq \sup\{\alpha, \beta, \gamma\}$, we obtain

$$\sup\{\sup\{\alpha, \beta\}, \gamma\} \leq \sup\{\alpha, \beta, \gamma\}.$$

The equality $\sup\{\sup\{\alpha, \beta\}, \gamma\} = \sup\{\alpha, \beta, \gamma\}$ follows from the antisymmetry of $\leq$. The proof of $\sup\{\alpha, \sup\{\beta, \gamma\}\} = \sup\{\alpha, \beta, \gamma\}$ is similar.

(2) Similar to (1).

(3) The proof is trivial. □

Generally speaking, a finite set S does not necessarily have the supremum or infimum in a poset. For example, given $P = \{a, b, c, d\}$, the relation $\leq$ on P is defined as $\leq = \{(a,a), (b,b), (c,c), (d,d), (a,c), (a,d), (b,c), (b,d)\}$. Then $(P, \leq)$ is a poset. For this poset, neither the supremum nor the infimum of $\{c, d\}$ exist. However, this is not the case in a lattice.

Definition 1.9. *A poset $(L, \leq)$ is called a lattice if $\sup\{\alpha, \beta\}$ and $\inf\{\alpha, \beta\}$ exist for any α, β in L.*

Clearly, $(\mathbb{R}, \leq)$ is a lattice. In this lattice, $\sup\{a, b\} = \max\{a, b\}$ and $\inf\{a, b\} = \min\{a, b\}$. By Example 1.6, $(\mathbb{Z}^+, |)$ is a lattice.

Example 1.7. *Consider the poset $(P(X), \subseteq)$ and $S = \{A, B\} \subseteq P(X)$. Then $\sup S = A \cup B$. Indeed, $A \subseteq A \cup B$ and $B \subseteq A \cup B$. So $A \cup B$ is an upper bound of S. Let C be an arbitrary upper bound of S, i.e. $A \subseteq C$ and $B \subseteq C$. Then $A \cup B \subseteq C$. Hence $A \cup B$ is the least upper bound of S or the supremum of S. Similarly, it can be shown that $\inf S = A \cap B$. As a consequence, $(P(X), \subseteq)$ is a lattice.*

Proposition 1.7. *In a lattice $(L, \leq)$, write $\alpha \vee \beta = \sup(\alpha, \beta)$ and $\alpha \wedge \beta = \inf(\alpha, \beta)$. Then the algebraic system $(L, \vee, \wedge)$ satisfies:*

(1) idempotency: $\forall \alpha \in L, \alpha \vee \alpha = \alpha$ and $\alpha \wedge \alpha = \alpha$;
(2) commutativity: $\forall \alpha, \beta \in L, \alpha \vee \beta = \beta \vee \alpha$ and $\alpha \wedge \beta = \beta \wedge \alpha$;
(3) associativity: $\forall \alpha, \beta, \gamma \in L$,

$$\alpha \vee (\beta \vee \gamma) = (\alpha \vee \beta) \vee \gamma \text{ and } \alpha \wedge (\beta \wedge \gamma) = (\alpha \wedge \beta) \wedge \gamma;$$

(4) absorption laws: $\forall \alpha, \beta \in L, \alpha \wedge (\alpha \vee \beta) = \alpha$ and $\alpha \vee (\alpha \wedge \beta) = \alpha$.

Proof. (1) and (2) are obvious.

(3) By Proposition 1.6(1),

$$\alpha \vee (\beta \vee \gamma) = \sup\{\alpha, \sup\{\beta, \gamma\}\} = \sup\{\sup\{\alpha, \beta\}, \gamma\} = (\alpha \vee \beta) \vee \gamma.$$

The proof of the other equality is similar.

(4) Observe that

$$\alpha \wedge (\alpha \vee \beta) = \inf\{\alpha, \sup\{\alpha, \beta\}\} \quad \text{and} \quad \alpha \vee (\alpha \wedge \beta) = \sup\{\alpha, \inf\{\alpha, \beta\}\}.$$

Since $\alpha \leq \sup\{\alpha, \beta\}$ and $\inf\{\alpha, \beta\} \leq \alpha$, we have the desired results by Proposition 1.6(3). □

Proposition 1.7 indicates that a lattice determines an algebraic system with four operation laws. The converse is also true, which is more precisely stated in the following proposition.

Proposition 1.8. *Assume that an algebraic system $(L, \vee, \wedge)$ with two algebraic operations $\vee, \wedge$ satisfies the idempotency, commutativity, associativity and absorption laws. Formulate an order relation $\leq$ by: $\alpha \leq \beta$ iff $\alpha \vee \beta = \beta$. Then $(L, \leq)$ is a lattice.*

Proof. Firstly, we show that $(L, \leq)$ is a poset.

Reflexivity follows from idempotency.

If $\alpha \leq \beta$, and $\beta \leq \gamma$, then $\alpha \vee \beta = \beta$ and $\beta \vee \gamma = \gamma$. Hence $\alpha \vee \gamma = \alpha \vee (\beta \vee \gamma) = (\alpha \vee \beta) \vee \gamma = \beta \vee \gamma = \gamma$ and thus $\alpha \leq \gamma$. The relation $\leq$ is transitive.

If $\alpha \leq \beta$ and $\beta \leq \alpha$, then $\alpha \vee \beta = \beta$ and $\beta \vee \alpha = \alpha$. By commutativity, $\alpha = \beta$. Hence the relation $\leq$ is antisymmetric.

Next we prove that $\sup\{\alpha, \beta\} = \alpha \vee \beta$ and $\inf\{\alpha, \beta\} = \alpha \wedge \beta$ for any α, β in L.

Indeed, $(\alpha \vee \beta) \vee \alpha = (\beta \vee \alpha) \vee \alpha = \beta \vee (\alpha \vee \alpha) = \beta \vee \alpha = \alpha \vee \beta$. Hence $\alpha \leq \alpha \vee \beta$. Similarly $\beta \leq \alpha \vee \beta$. Thus $\alpha \vee \beta$ is an upper bound of $\{\alpha, \beta\}$. Suppose γ is an upper bound of $\{\alpha, \beta\}$. Then $\alpha \leq \gamma$ and $\beta \leq \gamma$, i.e. $\alpha \vee \gamma = \beta \vee \gamma = \gamma$. Therefore, $(\alpha \vee \beta) \vee \gamma = \alpha \vee (\beta \vee \gamma) = \alpha \vee \gamma = \gamma$, which means that $\alpha \vee \beta \leq \gamma$. In summary, $\sup\{\alpha, \beta\} = \alpha \vee \beta$.

Furthermore, $\alpha \leq \beta$ iff $\alpha \wedge \beta = \alpha$. Indeed, $\alpha \leq \beta$ implies $\alpha \vee \beta = \beta$. Hence $\alpha \wedge \beta = \alpha \wedge (\alpha \vee \beta) = \alpha$ by the absorption law. Conversely, from $\alpha \wedge \beta = \alpha$, it follows that $\alpha \vee \beta = (\alpha \wedge \beta) \vee \beta = \beta$, i.e. $\alpha \leq \beta$.

Now $\inf\{\alpha, \beta\} = \alpha \wedge \beta$ can be verified in the same manner as $\sup\{\alpha, \beta\} = \alpha \vee \beta$. □

In view of Propositions 1.7 and 1.8, a lattice $(L, \leq)$ is also written as $(L, \vee, \wedge)$ if necessary. For example, $(P(X), \subseteq)$ and $(P(X), \cup, \cap)$ are considered to be identical.

1.5 Special Lattices

In this section, we impose some further restrictions on a lattice, which leads to some particular lattices.

Definition 1.10. *A lattice $(L, \vee, \wedge)$ is called distributive if it satisfies distributivity:* $\forall \alpha, \beta, \gamma \in L$,

$$\alpha \wedge (\beta \vee \gamma) = (\alpha \wedge \beta) \vee (\alpha \wedge \gamma), \quad \alpha \vee (\beta \wedge \gamma) = (\alpha \vee \beta) \wedge (\alpha \vee \gamma).$$

From Exercise 30 in this chapter, it is known that it suffices to require one of the above equalities to be valid for a lattice to be distributive.

Definition 1.11. *A lattice $(L, \leq)$ is said to be bounded if there exist the elements 1 and 0 in L such that $\forall \alpha \in L, 0 \leq \alpha \leq 1$.*

In a bounded lattice, 1 (resp. 0) is called the greatest (resp. least) element. For example, $(P(X), \subseteq)$ is a bounded lattice by Proposition 1.1 with the greatest element X and the least element $\emptyset$. When a lattice is regarded as an algebraic system, the boundedness means that there exist 0 and 1 in L such that $\forall \alpha \in L$,

$$0 \vee \alpha = \alpha, 1 \vee \alpha = 1, 0 \wedge \alpha = 0, 1 \wedge \alpha = \alpha.$$

For example, in the lattice $(P(X), \subseteq)$, $\forall A \in P(X)$,

$$\emptyset \cup A = A, \emptyset \cap A = \emptyset, X \cap A = A, X \cup A = X.$$

Definition 1.12. *A bounded lattice $(L, \leq)$ is called complemented if there exists a unary operation c in L such that $\forall \alpha \in L$, $\alpha \vee \alpha^c = 1$, $\alpha \wedge \alpha^c = 0$.*

In a complemented lattice, α^c is usually called a complement of α. It is worth noting that complement of an element is not necessarily unique. For example, consider the set $L = \{a, b, c, d, e\}$ with $\leq$ defined as

$$\{(a,a),(b,b),(c,c),(d,d),(e,e),(a,b),(a,c),(b,c),(a,d),(a,e),(d,c),(e,c),(d,e)\}.$$

Then the greatest and least element are c and a respectively. Since $b \vee e = b \vee d = c$ and $b \wedge e = b \wedge d = a$, both d and e are complements of b. However, it is not the case in a complemented distributive lattice (see Exercise 34 in this chapter).

Definition 1.13. *A complemented distributive lattice $(L, \leq)$ with at least two elements in L is called a Boolean lattice or Boolean algebra.*

For instance, $(P(X), \cup, \cap, c)$ is a Boolean algebra by Proposition 1.1.

Example 1.8. Let $L_0 = \{0, 1\}$. The operations $\vee, \wedge, c$ are defined in the following:

$$0 \vee 0 = 0, \quad 0 \vee 1 = 1 \vee 0 = 1 \vee 1 = 1;$$
$$1 \wedge 1 = 1, \quad 0 \wedge 0 = 0 \wedge 1 = 1 \wedge 0 = 0;$$
$$0^c = 1 \text{ and } 1^c = 0.$$

Then it can be easily checked that $(L_0, \vee, \wedge, c)$ is a Boolean algebra.

In a Boolean algebra, the complementation is a rather strong requirement. This notion is relevant to dealing with the structure of sets, which is the foundation of two-valued logic. However, it is irrelevant for the discussion of fuzzy set theory. To find a suitable structure for fuzzy sets, we need to introduce some additional concepts.

Definition 1.14. *If $(L, \leq)$ is a bounded distributive lattice and if there exists a unary operation c such that i) c is involutive, i.e. $\forall \alpha \in L$, $(\alpha^c)^c = \alpha$; ii) De Morgan laws are valid, i.e. $\forall \alpha, \beta \in L$, $(\alpha \vee \beta)^c = \alpha^c \wedge \beta^c$, $(\alpha \wedge \beta)^c = \alpha^c \vee \beta^c$, then $(L, \vee, \wedge, c)$ is called a Morgan algebra or soft algebra.*

1.5 Special Lattices

Example 1.9. *In the closed interval $[0,1]$, define $\vee$, $\wedge$ and c as follows: $\forall a,b \in [0,1]$, $a \vee b = \max\{a,b\}$, $a \wedge b = \min\{a,b\}$ and $a^c = 1-a$. Then it can be verified that $([0,1], \vee, \wedge, c)$ is a soft algebra. For instance, we prove one of De Morgan laws. Indeed, $\forall a,b \in [0,1]$, $(a \vee b)^c = 1 - \max\{a,b\} = \min\{1-a, 1-b\} = \min\{a^c, b^c\} = a^c \wedge b^c$.*

It is worth noting that, for the soft algebra $([0,1], \vee, \wedge, c)$, neither $\alpha \vee \alpha^c = 1$ nor $\alpha \wedge \alpha^c = 0$ is true. As a consequence, $([0,1], \vee, \wedge, c)$ is not a Boolean algebra, which indicates a soft algebra is not necessarily a Boolean algebra. As a matter of fact, the conditions of Boolean algebra are stronger than those of soft algebra. To verify this result, we first present a lemma.

Lemma 1.1. *Let $(L, \vee, \wedge, c)$ be a Boolean algebra. Then, for any $\alpha, \beta \in L$, $\alpha \vee \beta = 1$ and $\alpha \wedge \beta = 0$ imply $\beta = \alpha^c$.*

Proof. $\beta = 1 \wedge \beta = (\alpha^c \vee \alpha) \wedge \beta = (\alpha^c \wedge \beta) \vee (\alpha \wedge \beta) = \alpha^c \wedge \beta$.
$\alpha^c = \alpha^c \wedge 1 = \alpha^c \wedge (\alpha \vee \beta) = (\alpha^c \wedge \alpha) \vee (\alpha^c \wedge \beta) = \alpha^c \wedge \beta$.
Therefore $\beta = \alpha^c$. □

Proposition 1.9. *Every Boolean algebra is a soft algebra.*

Proof. It suffices to prove that involution and De Morgan laws are valid.
From $\alpha \vee \alpha^c = 1$ and $\alpha \wedge \alpha^c = 0$, it follows that $(\alpha^c)^c = \alpha$ by Lemma 1.1. In addition,

$$(\alpha \vee \beta) \vee (\alpha^c \wedge \beta^c) = \alpha \vee [(\beta \vee \alpha^c) \wedge (\beta \vee \beta^c)] = \alpha \vee [(\beta \vee \alpha^c) \wedge 1]$$
$$= \alpha \vee (\alpha^c \vee \beta) = (\alpha \vee \alpha^c) \vee \beta = 1,$$

$$(\alpha \vee \beta) \wedge (\alpha^c \wedge \beta^c) = (\alpha \wedge (\alpha^c \wedge \beta^c)) \vee (\beta \wedge \alpha^c \wedge \beta^c) = 0 \vee 0 = 0.$$

Applying Lemma 1.1 yields $(\alpha \vee \beta)^c = \alpha^c \wedge \beta^c$. Following a similar reasoning, $(\alpha \wedge \beta)^c = \alpha^c \vee \beta^c$ can be derived. □

Definition 1.15. *A lattice $(L, \leq)$ is called complete if every non-empty subset of L has an infimum and supremum.*

For a complete lattice $(L, \leq)$, the operations $\vee$ and $\wedge$ can be extended to infinitely many elements. Let $\alpha_i \in L$ ($i \in I$) with I an arbitrary index set. Define $\bigvee_{i \in I} \alpha_i$ as $\sup\{\alpha_i | i \in I\}$ and $\bigwedge_{i \in I} \alpha_i$ as $\inf\{\alpha_i | i \in I\}$.

A complete lattice $(L, \vee, \wedge)$ is called completely distributive if

$$\alpha \vee (\bigwedge_{i \in I} \alpha_i) = \bigwedge_{i \in I}(\alpha \vee \alpha_i)$$

$$\alpha \wedge (\bigvee_{i \in I} \alpha_i) = \bigvee_{i \in I}(\alpha \wedge \alpha_i),$$

where $\alpha, \alpha_i \in L$ with I an arbitrary index set.

Example 1.10. *Clearly, the soft algebra* $([0,1], \vee, \wedge)$ *is a complete lattice. Extend the operations* $\vee$ *and* $\wedge$ *as:*

$\forall a_i \in [0,1], \vee\{a_i | i \in I\} = \sup\{a_i | i \in I\}$;
$\forall a_i \in [0,1], \wedge\{a_i | i \in I\} = \inf\{a_i | i \in I\}$.

Now we prove that $([0,1], \vee, \wedge, c)$ *is completely distributive.*

Firstly,
$$\forall i \in I, \quad a \wedge a_i \leq a \wedge (\bigvee_{i \in I} a_i)$$

and thus
$$a \wedge (\bigvee_{i \in I} a_i) \geq \bigvee_{i \in I} (a \wedge a_i).$$

If $a \wedge (\bigvee_{i \in I} a_i) > \bigvee_{i \in I} (a \wedge a_i)$, *then there exists* b *such that*

$$a \wedge (\bigvee_{i \in I} a_i) > b > \bigvee_{i \in I} (a \wedge a_t).$$

On the one hand, $b > \bigvee_{i \in I}(a \wedge a_i)$ *implies that* $\forall i \in I$, $b > a \wedge a_i$. *On the other hand,* $b < a \wedge (\bigvee_{i \in I} a_i)$ *implies* $b < a$ *and* $b < \bigvee_{i \in I} a_i$. *Thus there exists* $i_0 \in I$ *such that* $b < a \wedge a_{i_0}$ *which is a contradiction. Therefore*

$$a \wedge (\bigvee_{i \in I} a_i) = \bigvee_{i \in I} (a \wedge a_i).$$

The proof of the other equality is similar. □

Definition 1.16. *Let* $(L, \vee, \wedge, c)$ *be a completely distributive soft algebra. If* L *is dense, i.e.* $\forall \alpha, \beta \in L$, $\alpha < \beta$ *implies that* $\exists \gamma \in L$, $\alpha < \gamma < \beta$, *then* $(L, \vee, \wedge, c)$ *is called a superior soft algebra.*

Clearly, $([0,1], \vee, \wedge, c)$ is a superior soft algebra. The properties of the algebraic system $([0,1], \vee, \wedge, c)$ as a superior soft algebra will be extensively employed in the coming chapters. Besides, we mention some additional equalities pertaining to $([0,1], \vee, \wedge, c)$, which are in store for future use. Let I be an arbitrary index set. If a_i and b_i ($i \in I$) are real numbers in $[0,1]$, then

(1) $(\bigwedge_{i \in I} a_i)^c = \bigvee_{i \in I} a_i^c$, $(\bigvee_{i \in I} a_i)^c = \bigwedge_{i \in I} a_i^c$;

(2) $(\bigwedge_{i \in I} a_i) \wedge (\bigwedge_{i \in I} b_i) = \bigwedge_{i \in I}(a_i \wedge b_i)$, $(\bigvee_{i \in I} a_i) \vee (\bigvee_{i \in I} b_i) = \bigvee_{i \in I}(a_i \vee b_i)$.

As a matter of fact, (1) and (2) are extensions of De Morgan laws and distributivity respectively. Their proofs are left to the reader.

Finally, we point out that $(P(X), \cup, \cap, c)$ is not a superior soft algebra. For example, let $X = \{a, b, c\}$, $A = \{a\}$ and $B = \{a, b\}$. Apparently, $A \subseteq B$ and $A \neq B$. However, there is no subset C of X such that $A \subseteq C \subseteq B$ and $A \neq C$, $C \neq B$. Hence, $P(X)$ is not dense, and thus not a superior soft algebra.

1.6 Exercises

1. Show that $\bigcup_{i \in I} A_i$ is the least set containing all A_i ($i \in I$) and $\bigcap_{i \in I} A_i$ is the greatest set contained in all A_i ($i \in I$).
2. Let A_i ($i = 1, 2, \cdots$) be subsets of $\mathbb{R}$, the set of real numbers, defined by $A_i = \{x | x \leq 1 - \frac{1}{i}\}$. Show that $\bigcup_{i=1}^{\infty} A_i =]-\infty, 1[$.
3. The difference $A \setminus B$ and symmetric difference $A \triangle B$ of two sets A and B are respectively defined by $A \setminus B = A \cap B^c$ and $A \triangle B = (A \setminus B) \cup (B \setminus A)$. Show that

   (i) $A^c \triangle B^c = A \triangle B$;
   (ii) $A \cap (B \triangle C) = (A \cap B) \triangle (A \cap C)$;
   (iii) $(A \triangle B) \triangle C = A \triangle (B \triangle C)$.

4. Show that if $A_1 \subseteq A_2 \subseteq A_3 \subseteq \cdots \subseteq A_n \subseteq \cdots$ and $B_1 \subseteq B_2 \subseteq B_3 \subseteq \cdots \subseteq B_n \subseteq \cdots$, then $(\bigcup_{n=1}^{\infty} A_n) \cap (\bigcup_{n=1}^{\infty} B_n) = \bigcup_{n=1}^{\infty} (A_n \cap B_n)$.
5. Find the characteristic function of $A \setminus B$ and $A \triangle B$ in Exercise 3 provided that the characteristic function of A and B are given.
6. Let A and B be two subsets of X. Show that the following equalities are valid for all $x \in X$:

   (i) $\chi_{A \cap B}(x) = \chi_A(x) \chi_B(x) = \max\{\chi_A(x) + \chi_B(x) - 1, 0\}$;
   (ii) $\chi_{A \cup B}(x) = \chi_A(x) + \chi_B(x) - \chi_A(x) \chi_B(x) = \min\{\chi_A(x) + \chi_B(x), 1\}$.

7. Let R, R_1 and R_2 be three relations from X to Y. Verify the following identities:

   (1) $\forall x \in X, y \in Y, xR = R^{-1}x$ and $Ry = yR^{-1}$;
   (2) $\forall x \in X, x(R_1 \cap R_2) = xR_1 \cap xR_2$ and $x(R_1 \cup R_2) = xR_1 \cup xR_2$;
   (3) $\forall y \in Y, (R_1 \cap R_2)y = R_1 y \cap R_2 y$ and $(R_1 \cup R_2)y = R_1 y \cup R_2 y$;
   (4) $\forall x \in X, y \in Y, xR^c = (xR)^c$ and $R^c y = (Ry)^c$.

8. Define the relations R_1 and R_2 on $A = \{0, 1, 2, 3\}$ by

   $$R_1 = \{(i,j) | j = i+1 \text{ or } j = \frac{i}{2}\}, \quad R_2 = \{(i,j) | i = j+1\}.$$

   Find $R_1 \cup R_2$, $R_1 \cap R_2$, R_1^{-1}, $R_1 \circ R_2$, $(R_1 \circ R_2) \circ R_1$, R_1^3.

9. If R_1 is a relation from A to B, R_2 and R_3 are relations from B to C, show that:

   (i) $R_1 \circ (R_2 \cup R_3) = (R_1 \circ R_2) \cup (R_1 \circ R_3)$;
   (ii) $R_1 \circ (R_2 \cap R_3) \subseteq (R_1 \circ R_2) \cap (R_1 \circ R_3)$. Give an example to illustrate that the inclusion cannot be replaced by the equality.

10. Verify the following identities:

    (i) $R_1 \triangleleft (R_2 \triangleright R_3) = (R_1 \triangleleft R_2) \triangleright R_3$;

(ii) $R_1 \triangleleft (R_2 \triangleleft R_3) = (R_1 \circ R_2) \triangleleft R_3$;
(iii) $R_1 \triangleright (R_2 \triangleright R_3) = R_1 \triangleright (R_2 \circ R_3)$.

11. Let $R = (r_{ik})_{l \times m}$ and $S = (s_{kj})_{m \times n}$. For $a, b \in [0, 1]$, write

$$a \to b = \begin{cases} 1 & a \leq b \\ 0 & \text{otherwise} \end{cases} \text{ and } a \leftrightarrow b = \begin{cases} 1 & a = b \\ 0 & \text{otherwise} \end{cases}.$$

Show that $(R \triangleleft S)_{ij} = \bigwedge_{k=1}^{m} (r_{ik} \to s_{kj})$, $(R \triangleright S)_{ij} = \bigwedge_{k=1}^{m} (s_{kj} \to r_{ik})$, and
$(R \square S)_{ij} = \bigwedge_{k=1}^{m} (r_{ik} \leftrightarrow s_{kj})$.

12. Show that R is transitive on X iff

$$\forall x, y, z \in X, \chi_R(x, z) \geq \min\{\chi_R(x, y), \chi_R(y, z)\}.$$

13. Let $X = \{1, 2, 3, 4\}$. A relation R on $P(X)$ is defined by: $\forall A, B \in P(X)$, ARB iff A and B have the same number of elements. Show that R is an equivalence relation. Find the quotient set of X by R.

14. Let R be a relation on A. Show that, for any positive integer k,
    (i) reflexivity of R implies reflexivity of R^k;
    (ii) symmetry of R implies symmetry of R^k;
    (iii) transitivity of R implies transitivity of R^k.

15. Let $R = \{(1, 2), (4, 3), (2, 2), (2, 1), (3, 1)\}$ be a relation on $A = \{1, 2, 3, 4\}$. Show that R is not transitive. Find a relation R' such that R' is the least transitive relation (in the sense of inclusion) containing R.

16. Let R_1 and R_2 be two equivalence relations on a finite set. Show that rank$(R_1 \cap R_2) \leq$ rank(R_1)rank(R_2), where rank$(R_1 \cap R_2)$, rank(R_1) and rank(R_2) stand for the rank of the matrices $R_1 \cap R_2$, R_1 and R_2 respectively.

17. Let R be a relation on A. $P = R \cap (R^{-1})^c$, $I = R \cap R^{-1}$. Show that
    (i) R is transitive iff P is transitive, I is transitive, $P \circ I \subseteq P$ and $I \circ P \subseteq P$;
    (ii) If R is transitive, then $\{x | \forall y \in A, y P^c x\} \neq \emptyset$.

18. Let f be a mapping from X to Y and $A \in P(X)$. Show that $\forall y \in f(X)$,
$$\chi_{f(A)}(y) = \bigvee_{f(x)=y} \chi_A(x).$$

19. Let R be a relation from X to Y such that $R^{-1} \circ R \subseteq I_Y$ and $R \circ R^{-1} \supseteq I_X$. Define the correspondence rule f_R by:

$$\forall x \in X, y = f_R(x) \Leftrightarrow (x, y) \in R.$$

Show that f_R is a mapping from X to Y.

1.6 Exercises

20. Define a mapping f from $\mathbb{Z}$ (the set of all integers) to $\mathbb{Z}$ by: $f(i) = 2i + 1$. Find $f \circ f$.
21. Give an example to illustrate that the composition of mappings does not satisfy commutativity.
22. Show that a mapping f from A to B is an injection iff $\forall X$ and $\forall \varphi, \psi$: $X \to A$, $f \circ \varphi = f \circ \psi$ implies $\varphi = \psi$.
23. Show that a mapping f from A to B is a surjection iff $\forall X$ and $\forall \varphi, \psi$: $B \to X$, $\varphi \circ f = \psi \circ f$ implies $\varphi = \psi$.
24. Let f be a mapping from X to Y. Show that
    (1) f is injective iff there exists a mapping g from Y to X such that $g \circ f = I_X$;
    (2) f is surjective iff there exists a mapping g from Y to X such that $f \circ g = I_Y$;
    (3) f is bijective iff there exists a mapping g from Y to X such that $g \circ f = I_X$ and $f \circ g = I_Y$.
25. Let f be a mapping from X to Y. Then f would induce a relation R_f from X to Y, which is defined by $R_f = \{(x, f(x)) | x \in X\}$. Show that:
    (i) The induced relation R_f satisfies that $R_f^{-1} \circ R_f \subseteq I_Y$ and $R_f \circ R_f^{-1} \supseteq I_X$;
    (ii) If f is injective, then $R_f \circ R_f^{-1} = I_X$;
    (iii) If f is surjective, then $R_f^{-1} \circ R_f = I_Y$.
26. Let f be a mapping from a finite set X to an arbitrary set Y. Define an equivalence relation R_f on X by: $x_1 R_f x_2$ iff $f(x_1) = f(x_2)$. Suppose that $f(X) = \{y_1, y_2, \cdots, y_n\}$ and $X/R_f = \{A_1, A_2, \cdots, A_n\}$. Show that if one makes the convention: $y0 = 0$ and $y1 = y$ for any $y \in Y$, then

$$\forall x \in X, f(x) = y_1 \chi_{A_1}(x) + y_2 \chi_{A_1}(x) + \cdots + y_n \chi_{A_n}(x).$$

27. Let $\mathbb{Z}$ denote the set of all integers and $\mathbb{N}$ the set of all natural numbers. Show that there exists a bijective mapping from $\mathbb{Z}$ to $\mathbb{N}$. Let the symbols $+$ and $\cdot$ denote the addition and multiplication of integers respectively. Show that the algebraic systems $(\mathbb{Z}, +)$ and $(\mathbb{N}, \cdot)$ are not isomorphic.
28. Show that if an algebraic system $(L, \vee, \wedge)$ satisfies the absorption laws, then it also satisfies idempotency.
29. Show that a lattice $(L, \leq)$ has the following properties:
    (i) $\beta \leq \gamma \Rightarrow \alpha \wedge \beta \leq \alpha \wedge \gamma$ and $\alpha \vee \beta \leq \alpha \vee \gamma$;
    (ii)
    $$\alpha \vee (\beta \wedge \gamma) \leq (\alpha \vee \beta) \wedge (\alpha \vee \gamma)$$
    $$\alpha \wedge (\beta \vee \gamma) \geq (\alpha \wedge \beta) \vee (\alpha \wedge \gamma);$$
    (iii) $\alpha \leq \gamma \Leftrightarrow \alpha \vee (\beta \wedge \gamma) \leq (\alpha \vee \beta) \wedge \gamma$.

30. Let $(L, \vee, \wedge)$ be a lattice. Show that if one of the following equalities is valid

$$a \wedge (b \vee c) = (a \wedge b) \vee (a \wedge c)$$
$$a \vee (b \wedge c) = (a \vee b) \wedge (a \vee c)$$

then the other one is also valid.

31. Let $(L_1, \leq_1)$ and $(L_2, \leq_2)$ be two lattices and f be a bijection from L_1 to L_2. Show that f is an isomorphism between $(L_1, \leq_1)$ and $(L_2, \leq_2)$ iff $\forall a, b \in L_1,\ a \leq_1 b \Leftrightarrow f(a) \leq_2 f(b)$.

32. Show that in a distributive lattice $(L, \vee, \wedge)$, $\alpha \wedge \beta = \alpha \wedge \gamma$ and $\alpha \vee \beta = \alpha \vee \gamma$ imply $\beta = \gamma$.

33. The relation $\leq$ on $L = [0,1]^n$ is defined by

$$\forall a = (a_1, a_2, \ldots, a_n) \in L, b = (b_1, b_2, \ldots, b_n) \in L,$$

$$a \leq b \Leftrightarrow a_i \leq b_i (i = 1, 2, \ldots, n).$$

Show that $(L, \leq)$ is a completely distributive lattice.

34. Show that the complement of every element is unique in a complemented distributive lattice.

35. Show that a bounded lattice $(L, \leq)$ is complete iff every non-empty subset of L has an infimum or supremum.

36. Show that if $(L, \leq)$ is a dense complete lattice, then $\sup\{\alpha | \alpha < \beta\} = \beta$ and $\inf\{\alpha | \alpha > \beta\} = \beta$.

37. Let $(L_1, \vee_1, \wedge_1, c)$ be a Boolean algebra and f a surjective homomorphism from $(L_1, \vee_1, \wedge_1, c)$ to $(L_2, \vee_2, \wedge_2, c)$. Show that $(L_2, \vee_2, \wedge_2, c)$ is also a Boolean algebra.

38. Let $(L, \vee, \wedge, c)$ be an algebraic system satisfying commutativity, distributivity. If, furthermore, there exist $0, 1$ in L such that $0 \vee \alpha = \alpha$, $\alpha \wedge 1 = \alpha$, $\alpha \vee \alpha^c = 1$ and $\alpha \wedge \alpha^c = 0$, show that $(L, \vee, \wedge, c)$ is a Boolean algebra.

Chapter 2
Basics of Fuzzy Sets

In this chapter, we focus on the introduction of fundamentals in fuzzy set theory, including some set-theoretic operations and their extensions, the decomposition of a fuzzy set, and mathematical representations of fuzzy sets in terms of a nest of sets. Towards the end of the chapter, fuzzy sets taking values in $[0,1]$ are extended to those on a lattice and a similar investigation is carried out.

2.1 Fuzzy Sets and Their Set-Theoretic Operations

According to Cantor, a set consists of some elements which are definite. In other words, for a given element, whether it belongs to the set or not should be clear. As a consequence, a set can only be employed to describe a concept which is crisply defined. For example, a collection of cities with the population more than 5 millions forms a set since we can judge that a given city is in this set or not without vagueness. In traditional mathematics, all the involved concepts ranging all the way from the complex numbers and matrices to geometric transformations and algebraic structures are in this category. However, in the real world, mankind often uses concepts which are quite vague. For example, we say that a man is young or middle-aged, an object is expensive or cheap, a tomato is red and mature, a number is large or small, a car is slow or fast and so on. Let us take *young* as an illustration. Suppose A is a 20-year-old man. Maybe you think A is certainly young. Now comes a man B only one day elder than A. Of course, B is still young. Then how about a man only one day elder than B. Continuing in this way, you will find it difficult to determine an exact age beyond which a man will be middle-aged. As a matter of fact, there is no sharp line between young and middle-aged. The transition from one concept to the other is gradual. This gradualness results in the vagueness of the concept *young*, which in return makes the boundary of the set of all young men unclear. In 1965, Zadeh introduced the concept of fuzzy sets just in order to represent this class of sets. In his seminal paper *Fuzzy Sets* [171], Zadeh assigns a number to every element in the universe, which indicates the degree (grade) to which the element belongs to a fuzzy set. In this interpretation, everybody has a degree to which he/she is young (eventually the degree may be 0 or 1). The people with different ages may have different degrees. To formulate this concept of fuzzy set mathematically, we present the following definition.

Definition 2.1. *Let X be the universe. A mapping $A : X \to [0,1]$ is called a fuzzy set on X. The value $A(x)$ of A at $x \in X$ stands for the degree of membership of x in A. The set of all fuzzy sets on X will be denoted by $F(X)$.*

$A(x) = 1$ means full membership, $A(x) = 0$ means non-membership and intermediate values between 0 and 1 mean partial membership. $A(x)$ is referred to as a membership function as x varies in X.

Remark 2.1. *It follows from the isomorphism between $(P(X), \cup, \cap, c)$ and $(Ch(X), \vee, \wedge, c)$ that every subset of X may be regarded as a mapping from X to $\{0, 1\}$. In this sense an ordinary set is also a fuzzy set, whose membership function is just its characteristic function. Accordingly we shall identify the membership degree $A(x)$ with the value $\chi_A(x)$ of the characteristic function χ_A at x when A is an ordinary set. For the two extreme cases $\emptyset$ (the empty set) and X (the entire set), the membership functions are defined by $\forall x \in X$, $\emptyset(x) = 0$ and $X(x) = 1$, respectively. In contrast with fuzzy sets, ordinary sets are sometimes termed by crisp sets in this book.*

Example 2.1. *Let O denote old and Y denote young. We limit the scope of age to $X = [0, 100]$. Then both O and Y are fuzzy sets that are respectively defined by Zadeh as follows:*

$$O(x) = \begin{cases} [1 + (\frac{x-50}{5})^{-2}]^{-1} & \text{if } 50 < x \leq 100; \\ 0 & \text{otherwise}. \end{cases}$$

$$Y(x) = \begin{cases} [1 + (\frac{x-25}{5})^2]^{-1} & \text{if } 25 < x \leq 100 \\ 1 & \text{otherwise} \end{cases}$$

For instance, $O(60) = 0.8$ and $Y(30) = 0.5$.

Example 2.2. *As known to us, all the involved quantities are precise in traditional mathematics. With fuzzy sets, we can model the so-called fuzzy data. For instance, the fuzzy datum $A=$"around 1" may be represented by: $\forall x \in \mathbb{R}$,*

$$A(x) = \begin{cases} x & 0 \leq x \leq 1 \\ 2 - x & 0 \leq x \leq 2 \\ 0 & \text{otherwise}. \end{cases}$$

In the case of infinite universe, a fuzzy set may be represented by its membership function as in Example 2.1. If the universe is finite, say, $X = \{x_1, x_2, \ldots, x_n\}$, the fuzzy set A on X is represented by

2.1 Fuzzy Sets and Their Set-Theoretic Operations

$$A = A(x_1)/x_1 + A(x_2)/x_2 + \cdots + A(x_n)/x_n.$$

For example, the fuzzy set $S = \textit{several}$ on $X = \{1, 2, \cdots, 10\}$ may be written as:

$$S = 0/1 + 0.6/2 + 1/3 + 1/4 + 1/5 + 0.9/6 + 0.8/7 + 0.7/8 + 0.6/9 + 0/10.$$

For the sake of conciseness, the terms with degree 0, e.g. the terms $0/1$, $0/10$ in S, are dropped. As a result,

$$S = 0.6/2 + 1/3 + 1/4 + 1/5 + 0.9/6 + 0.8/7 + 0.7/8 + 0.6/9.$$

Importantly, the choice of a membership function is context-dependent. It is clearly different that the temperature of a steel-smelting furnace is high and the temperature of a human body is high. Even in a same context, the choice is dependent on the observer. It is certainly different from Zadeh's if you form the membership function of the fuzzy concept *young*.

Next we introduce some set-theoretic operations of fuzzy sets formulated by Zadeh. Let A and B be two fuzzy sets on X.

The union $A \cup B$ of A and B is defined by

$$\forall x \in X, (A \cup B)(x) = \max(A(x), B(x)) (\text{or simply } A(x) \vee B(x));$$

The intersection $A \cap B$ of A and B is defined by

$$\forall x \in X, (A \cap B)(x) = \min(A(x), B(x)) (\text{or } A(x) \wedge B(x));$$

The complement A^c of A is defined by

$$\forall x \in X, A^c(x) = 1 - A(x).$$

Remark 2.2. *As in crisp case, the union (intersection) of fuzzy sets A and B represents "A or (resp. and) B", and the complement of A means "not A".*

Example 2.3. *Let $X = \{1, 2, \cdots, 10\}$.*

$A = \textit{small} = 1/1 + 0.8/2 + 0.6/3 + 0.4/4 + 0.2/5,$
$B = \textit{large} = 0.2/4 + 0.4/5 + 0.6/6 + 0.8/7 + 1/8 + 1/9 + 1/10.$

Then

not small

$$= A^c = 0.2/2 + 0.4/3 + 0.6/4 + 0.8/5 + 1/6 + 1/7 + 1/8 + 1/9 + 1/10,$$

not large

$$= B^c = 1/1 + 1/2 + 1/3 + 0.8/4 + 0.6/5 + 0.4/6 + 0.2/7,$$

not small and not large

$$= A^c \cap B^c = 0.2/2 + 0.4/3 + 0.6/4 + 0.6/5 + 0.4/6 + 0.2/7.$$

If $\forall x \in X$, $A(x) \leq B(x)$, then we call that A is a subset of B or A is contained in B, denoted by $A \subseteq B$.

If $\forall x \in X$, $A(x) = B(x)$, then A and B are called equal, denoted by $A = B$.

Obviously, $A = B$ iff $A \subseteq B$ and $B \subseteq A$.

If $A \neq \emptyset$, $A \subseteq B$ and $\exists x \in X$ such that $A(x) < B(x)$, then we say that A is properly contained in B, denoted by $A \subset B$. It follows immediately from the definitions that $\forall A, B, C, D \in F(X)$,

(1) $A \cap B \subseteq A$ and $A \subseteq A \cup B$;
(2) $A \subseteq B \iff A \cup B = B \iff A \cap B = A$;
(3) $A \subseteq B$ and $C \subseteq D \Rightarrow A \cup C \subseteq B \cup D$ and $A \cap C \subseteq B \cap D$;
(4) $A \subseteq B \Rightarrow B^c \subseteq A^c$.

In addition, we have the following important conclusion concerning the fuzzy set-theoretic operations.

Theorem 2.1. $(F(X), \cup, \cap, c)$ *is a soft algebra, i.e.* $F(X)$ *satisfies:* $\forall A, B, C \in F(X)$,

(1) idempotency: $A \cup A = A$, $A \cap A = A$;
(2) commutativity: $A \cup B = B \cup A$, $A \cap B = B \cap A$;
(3) associativity: $(A \cup B) \cup C = A \cup (B \cup C)$, $(A \cap B) \cap C = A \cap (B \cap C)$;
(4) absorption laws: $A \cup (A \cap B) = A$, $A \cap (A \cup B) = A$;
(5) distributivity: $A \cup (B \cap C) = (A \cup B) \cap (A \cup C)$, $A \cap (B \cup C) = (A \cap B) \cup (A \cap C)$;
(6) the existence of the greatest and least element: $\emptyset \subseteq A \subseteq X$.
(7) involution: $(A^c)^c = A$;
(8) De Morgan laws: $(A \cup B)^c = A^c \cap B^c$, $(A \cap B)^c = A^c \cup B^c$.

Proof. We prove, for instance, the De Morgan law: $(A \cup B)^c = A^c \cap B^c$. Indeed, $\forall x \in X$,

$$\begin{aligned}(A \cup B)^c(x) &= 1 - (A \cup B)(x) \\ &= 1 - (A(x) \vee B(x)) \\ &= (1 - A(x)) \wedge (1 - B(x)) \\ &= A^c(x) \wedge B^c(x) \\ &= (A^c \cap B^c)(x).\end{aligned}$$

Therefore $(A \cup B)^c = A^c \cap B^c$. □

From the above proof, we see that properties of $(F(X), \cup, \cap, c)$ are largely dependent on properties of $([0,1], \vee, \wedge, c)$ since the set-theoretic operations are defined pointwise. In this sense, $[0,1]$ is regarded as the underlying structure set of $F(X)$. As a result, it is not strange that $(F(X), \cup, \cap, c)$ has the same algebraic structure as $([0,1], \vee, \wedge, c)$. The partial order relation $\leq$ in the soft algebra $(F(X), \cup, \cap, c)$ is $\subseteq$. The proof of this assertion is left to the reader as an exercise (Exercise 4).

Like $([0,1], \vee, \wedge, c)$, $(F(X), \cup, \cap, c)$ is not a Boolean algebra since it is not complemented, i.e. $A \cap A^c = \emptyset$ and $A \cup A^c = X$ do not hold generally. To illustrate this point, consider the fuzzy set A defined by $\forall x \in X$, $A(x) = 0.5$. Then $\forall x \in X$, $(A \cap A^c)(x) = (A \cup A^c)(x) = 0.5$ while $\emptyset(x) = 0$ and $X(x) = 1$. Consequently, $A \cap A^c \neq \emptyset$ and $A \cup A^c \neq X$, which indicates that neither the law of contradiction nor the law of excluded middle hold. It is quite natural considering that these two laws are the logical foundation of traditional mathematics. In this sense, the emergence of fuzzy sets gives birth to a completely new logic – fuzzy logic, and hence to a completely new mathematics – mathematics of fuzziness.

In order to examine whether $(F(X), \cup, \cap, c)$ is a superior soft algebra or not, we extend the union and intersection operation.

For $A_i \in F(X) (i \in I)$ with an arbitrary index set I, $\bigcup_{i \in I} A_i$ is defined by

$$\forall x \in X, (\bigcup_{i \in I} A_i)(x) = \sup\{A_i(x) | i \in I\} (\text{ or } \bigvee_{i \in I} A_i(x))$$

and $\bigcap_{i \in I} A_i$ is defined by

$$\forall x \in X, (\bigcap_{i \in I} A_i)(x) = \inf\{A_i(x) | i \in I\} (\text{ or } \bigwedge_{i \in I} A_i(x)).$$

Theorem 2.2. $(F(X), \cup, \cap, c)$ *is a superior soft algebra.*

Proof. It can be deduced that $(F(X), \cup, \cap, c)$ is completely distributive from the complete distributivity of $([0,1], \vee, \wedge, c)$. What is left is to examine the density of the algebraic system $(F(X), \cup, \cap, c)$. Since $\leq$ is just $\subseteq$ in the soft algebra $(F(X), \cup, \cap, c)$, we assume that $A \subseteq B$ and $A \neq B$, i.e. $\forall x \in X$, $A(x) \leq B(x)$ and there exists an $x_0 \in X$ such that $A(x_0) < B(x_0)$. Define the fuzzy set C on X by $C(x) = \frac{1}{2}(A(x) + B(x)) (\forall x \in X)$. It is easily verified that $A(x) \leq C(x) \leq B(x)$ ($\forall x \in X$) and $A(x_0) < C(x_0) < B(x_0)$, whence $A \subseteq C \subseteq B$ and $A \neq C \neq B$, which completes the proof. $\square$

To conclude this section, we introduce some concepts and notations concerning a fuzzy set A on X.

(1) The set $\{x | A(x) = 1\}$ is said to be the kernel of A, denoted by $\ker(A)$;
(2) The set $\{x | A(x) > 0\}$ is called the support of A, denoted by $\mathrm{supp}(A)$;
(3) The number $\bigvee_{x \in X} A(x)$ is called the height of A, denoted by $\mathrm{hgt}(A)$, and the number $\bigwedge_{x \in X} A(x)$ is referred to as the plinth of A, denoted by $\mathrm{plt}(A)$.
(4) If $\ker(A) \neq \emptyset$, then A is called a normal fuzzy set.

For instance, in Example 2.1, supp(O) =$]50,100]$, hgt(O) = $\frac{100}{101}$, plt(O)=0 ker$(O) = \emptyset$ and supp$(Y) = [0,100]$, hgt(Y)=1, plt$(Y) = \frac{1}{226}$, ker$(Y) = [0,25]$. Hence Y is normal and O is not. Clearly, hgt$(A) = 1$ for every normal fuzzy set A.

2.2 General Fuzzy Logic Connectives

As we know in Section 1.1, $\forall x, \chi_{A \cap B}(x) = \min(\chi_A(x), \chi_B(x))$ which justifies Zadeh's use of the minimum operator in formulating the intersection of two fuzzy sets. Meanwhile, it is seen from Exercise 6 in Chapter 1 that $\chi_{A \cap B}(x) = \chi_A(x)\chi_B(x) = \max\{\chi_A(x)+\chi_B(x)-1, 0\}$. Hence it is also reasonable to define the intersection of A and B in $F(X)$ by $(A \cap B)(x) = A(x)B(x)$ or by $(A \cap B)(x) = \max\{A(x)+B(x)-1, 0\}$ if we consider the intersection of fuzzy sets as an extension of the intersection of crisp sets. The similar argument exists for the definition of the complement and union. In other words, to extend operations of crisp sets to the fuzzy case, there may be multiple alternative ways. The definition in the previous section is just one of them. More generally, the operation of intersection, union and complement can be formulated by means of the so-called t−norms, t−conorms, and fuzzy negations, respectively, together with fuzzy implications and fuzzy equivalencies, which will be briefly introduced in this section.

2.2.1 Fuzzy Negations

Definition 2.2. *If $n : [0,1] \to [0,1]$ is decreasing and satisfies the boundary conditions: $n(0) = 1, n(1) = 0$, then n is called a (fuzzy) negation.*

If we define n by $\forall x \in [0,1], n(x) = 1 - x$, then n is a negation, which is called the standard negation.

Example 2.4. *The mapping $n_i : [0,1] \to [0,1]$ defined by*

$$\forall x \in [0,1], n_i(x) = \begin{cases} 1 & x = 0 \\ 0 & x > 0 \end{cases},$$

is a negation, which is called the intuitionistic negation; and $n_{d_i}(x) = 1 - n_i(1-x)$, i.e.

$$n_{d_i}(x) = \begin{cases} 0 & x = 1 \\ 1 & x < 1 \end{cases}$$

is also a negation, which is called the dual intuitionistic negation.

Clearly, $\forall x \in [0,1], n_i(x) \leq n(x) \leq n_{d_i}(x)$ holds for any negation n, i.e. n_i is the minimum negation and n_{d_i} is the maximum negation.

Definition 2.3. *A strictly decreasing continuous negation is called a strict negation. A strict negation n is called a strong negation if it satisfies the involution: $\forall x \in [0,1], n(n(x)) = x$.*

2.2 General Fuzzy Logic Connectives

Neither n_i nor n_{d_i} is strict while the standard negation is a strong negation.

Example 2.5. *Define n by $\forall x \in [0,1], n(x) = 1 - x^2$. Then n is a strict negation, which is not strong.*

Definition 2.4. *Let $\varphi : [a,b] \to [a,b]$ be a strictly increasing and continuous function. If φ satisfies $\varphi(a) = a$ and $\varphi(b) = b$, then φ is called an automorphism on $[a,b]$.*

For example, both $\varphi_1(x) = x$ and $\varphi_2(x) = x^2$ are automorphisms on $[0,1]$ and $\varphi_3(x) = x^2 + x - 1/4$ is an automorphism on $[-1/2, 1/2]$.

Lemma 2.1. *Let n_1 and n_2 be two strict negations. Then there exist two automorphisms φ and ψ on $[0,1]$ such that $n_2 = \psi \circ n_1 \circ \varphi$.*

Proof. Let $s_1, s_2 \in [0,1]$ be such that $n_1(s_1) = s_1$, $n_2(s_2) = s_2$. Since $n_1(0) = n_2(0) = 1$, we have $s_1 \neq 0$ and $s_2 \neq 0$. Let $t = \dfrac{s_2}{s_1}$. Define φ and ψ by

$$\forall x \in [0,1], \varphi(x) = \begin{cases} \dfrac{x}{t} & x \leq s_2 \\ n_1^{-1}\left(\dfrac{n_2(x)}{t}\right) & x > s_2 \end{cases}$$

$$\psi(x) = \begin{cases} tx & x \leq s_1 \\ n_2[tn_1^{-1}(x)] & x > s_1 \end{cases}.$$

It is easily checked that φ and ψ are automorphisms on $[0,1]$.

When $x < s_2$, $\dfrac{x}{t} < \dfrac{s_2}{t} = s_1$, and thus $n_1(\dfrac{x}{t}) > n(s_1) = s_1$.

$$(\psi \circ n_1 \circ \varphi)(x) = \psi(n_1(\varphi(x))) = \psi(n_1(\dfrac{x}{t})) = n_2[tn_1^{-1}(n_1(\dfrac{x}{t}))] = n_2(x).$$

When $x \geq s_2$, $n_2(x) \leq n_2(s_2) = s_2 = s_1 t$, and thus $\dfrac{n_2(x)}{t} \leq s_1$.

$$(\psi \circ n_1 \circ \varphi)(x) = \psi[n_1(n_1^{-1}(\dfrac{n_2(x)}{t}))] = \psi(\dfrac{n_2(x)}{t}) = t \cdot \dfrac{n_2(x)}{t} = n_2(x).$$

Consequently, $n_2 = \psi \circ n_1 \circ \varphi$. □

It is easily checked that the converse of Lemma 2.1 is true, i.e. if n_1 is a strict negation and there exist two automorphisms φ and ψ on $[0,1]$ such that $n_2 = \psi \circ n_1 \circ \varphi$, then n_2 is a strict negation as well.

Theorem 2.3. *(Representation theorem of a strict negation) The negation $n : [0,1] \to [0,1]$ is strict iff there exist two automorphisms φ and ψ on $[0,1]$ such that*

$$n(x) = \psi(1 - \varphi(x))(\forall x \in [0,1]).$$

Proof. Let $n'(x) = 1 - x$. Then n' is a strict negation. From Lemma 2.1, we know that there exist two automorphisms φ and ψ on [0,1] such that $n = \psi \circ n' \circ \varphi$, i.e.

$$n(x) = \psi(1 - \varphi(x))(\forall x \in [0,1]).$$

The reverse implication is straightforward. □

Similar to the proof of Lemma 2.1, we can show that:

Lemma 2.2. *Let N_1 and N_2 be two strong negations. Then there exists an automorphism φ on $[0,1]$ such that $N_2 = \varphi^{-1} \circ N_1 \circ \varphi$.*

By applying Lemma 2.2, the following theorem can be proved (we leave the proof of Lemma 2.2 and Theorem 2.4 as an exercise (Exercise 13)).

Theorem 2.4. *(Representation theorem of a strong negation) The mapping $N : [0,1] \to [0,1]$ is a strong negation iff there exists an automorphism φ on [0,1] such that*

$$N(x) = \varphi^{-1}(1 - \varphi(x))(\forall x \in [0,1]).$$

It follows that every strong negation N can be expressed as $N(x) = \varphi^{-1}(1 - \varphi(x))$, where φ is an automorphism on $[0,1]$, which is called a generator of N. The strong negation N with the generator φ will be denoted by N_φ. Generally speaking, generator of a strong negation is not unique. For example, both $\varphi_1(x) = x$ and

$$\varphi_2(x) = \begin{cases} \sqrt{\frac{x}{2}} & x < 0.5 \\ 1 - \sqrt{\frac{1-x}{2}} & x \geq 0.5 \end{cases}$$

are generators of the standard negation $N(x) = 1 - x$.

2.2.2 Triangular Norms and Conorms

Definition 2.5. *A mapping T from $[0,1] \times [0,1]$ to $[0,1]$ is called a triangular norm (t-norm) or a conjunction, if it satisfies:*

(1) *symmetry:* $T(x,y) = T(y,x)$ *whenever* $x,y \in [0,1]$;
(2) *monotonicity:* $T(x_1, y_1) \leq T(x_2, y_2)$ *whenever* $x_1 \leq x_2$ *and* $y_1 \leq y_2$;
(3) *associativity:* $T(T(x,y), z) = T(x, T(y,z))$ *whenever* $x, y, z \in [0,1]$;
(4) *boundary condition:* $T(1,x) = x$ *whenever* $x \in [0,1]$.

We list some popular t-norms here:

$T_{\min}(x,y) = x \wedge y$
$T_L(x,y) = \max(0, x + y - 1)$ (Lukasiewicz t-norm)
$T_0(x,y) = \begin{cases} x & \text{if } y = 1; \\ y & \text{if } x = 1; \\ 0 & \text{otherwise} \end{cases}$
$T_\pi(x,y) = xy$.

2.2 General Fuzzy Logic Connectives

Let φ be an automorphism on $[0,1]$ and T a t-norm. Define T^φ by:

$$\forall x, y \in [0,1], T^\varphi(x,y) = \varphi^{-1}(T(\varphi(x), \varphi(y))).$$

It is easily checked that T^φ is a t-norm as well which is called the φ-transform of T. For instance, the φ-transform of T_L is

$$T_L^\varphi(x,y) = \varphi^{-1}(0 \vee (\varphi(x) + \varphi(y) - 1)).$$

In essence, a $t-$norm is a function of two variables. If the function is continuous, then we say that the $t-$norm is continuous. If the partial mapping $T(x, \cdot)$ is left continuous, then we say that T is left continuous. Considering the symmetry of a t-norm, the partial mapping $T(\cdot, y)$ is also left continuous if T is left continuous. In addition, the order relations between $t-$norms are naturally based on the comparisons between functions. For instance, we mean that $\forall x, y \in [0,1], T_1(x,y) \leq T_2(x,y)$ by $T_1 \leq T_2$.

Proposition 2.1. *(1)* $T_0 \leq T_L \leq T_\pi \leq T_{\min}$; *(2)* $T_0 \leq T \leq T_{\min}$ *holds for any $t-$norm T.*

Proof. (1) The proof is left to the reader.
(2) By monotonicity and boundary condition,

$$\forall x, y \in [0,1], T(x,y) \leq T(x,1) = x.$$

Similarly $T(x,y) \leq y$. Hence, $T(x,y) \leq x \wedge y = T_{\min}(x,y)$.

If $x = 1$ or $y = 1$, $T(x,y) = T_0(x,y)$ by boundary condition and symmetry, otherwise, $T_0(x,y) = 0 \leq T(x,y)$. We always have $T_0 \leq T$. $\square$

So the set of all t-norms is bounded with the greatest t-norm $T_{\min}$ and the least t-norm T_0.

Proposition 2.2. *If a $t-$norm satisfies idempotency: $T(x,x) = x(\forall x \in [0,1])$, then $T = T_{\min}$.*

Proof. By idempotency, monotonicity and Proposition 2.1(2), we have

$$\forall x, y \in [0,1], x \wedge y = T(x \wedge y, x \wedge y) \leq T(x,y) \leq x \wedge y.$$

Hence $T(x,y) = x \wedge y$. $\square$

Definition 2.6. *A mapping S from $[0,1] \times [0,1]$ to $[0,1]$ is called a triangular conorm (t−conorm) or a disjunction, if it satisfies:*

(1) symmetry: $S(x,y) = S(y,x)$ whenever $x, y \in [0,1]$;
(2) monotonicity: $S(x_1, y_1) \leq S(x_2, y_2)$ whenever $x_1 \leq x_2$ and $y_1 \leq y_2$;
(3) associativity: $S(S(x,y), z) = S(x, S(y,z))$ whenever $x, y, z \in [0,1]$;
(4) boundary condition: $S(0,x) = x$ whenever $x \in [0,1]$.

Remark 2.3. *Let T be a $t-$norm and S be a $t-$conorm. From an algebraic view, both $([0,1], T)$ and $([0,1], S)$ are semigroups with identities 1 and 0 respectively, and thus they are monoids.*

Let φ be an automorphism on $[0,1]$ and S a t-conorm. Define S^φ by:

$$\forall x, y \in [0,1], S^\varphi(x,y) = \varphi^{-1}(S(\varphi(x), \varphi(y))).$$

It is easily checked that S^φ is a t-conorm as well which is called the φ-transform of S. For instance, the φ-transform of S_L is

$$\forall x, y \in [0,1], S_L^\varphi(x,y) = \varphi^{-1}(0 \wedge (\varphi(x) + \varphi(y))).$$

Let T be a t-norm and n a strict negation. Define S by:

$$\forall x, y \in [0,1], S(x,y) = n^{-1}(T(n(x), n(y))).$$

Then it is easily checked that S is a t-conorm. Particularly, if n is the standard negation, then $S(x,y) = 1 - T(1-x, 1-y)$ is a t-conorm. With this result, we can derive the corresponding t-conorms from the popular t-norms mentioned above.

If $T = T_{\min}$, then we have

$$S(x,y) = 1 - T(1-x, 1-y) = 1 - ((1-x) \wedge (1-y)) = \max\{x,y\},$$

which will be denoted by $S_{\max}(x,y)$.

If $T = T_L$, then we obtain $S_L(x,y) = \min(1, x+y)$ (Lukasiewicz t-conorm).

If $T = T_0$, then we have $S_0(x,y) = \begin{cases} x & \text{if } y = 0; \\ y & \text{if } x = 0; \\ 1 & \text{otherwise} \end{cases}$

If $T = T_\pi$, then we have $S_\pi(x,y) = x + y - xy$.

Similar to the proofs of the results pertinent to t-norms, we can show that

(1) $S_0 \geq S_L \geq S_\pi \geq S_{\max}$;
(2) $S_0 \geq S \geq S_{\max}$ holds for every t-conorm S;
(3) If a t-conorm satisfies idempotency: $\forall x \in [0,1], S(x,x) = x$, then $S = S_{\max}$.

Proposition 2.3. *Let T and S be a t-norm and a t-conorm respectively.*

(1) *If T and S satisfy the absorption law:* $\forall x, y \in [0,1], T(S(x,y), x) = x$, *then* $T = T_{\min}$.
(2) *If T and S satisfy the absorption law:* $\forall x, y \in [0,1], S(T(x,y), x) = x$, *then* $S = S_{\max}$.

Proof. (1) Letting $y = 0$ yields $T(S(x,0), x) = x$, i.e. $T(x,x) = x (\forall x \in [0,1])$. The desired result follows from Proposition 2.2. As for the proof of (2), let $y = 1$. □

Proposition 2.4. *Let T and S be a t-norm and a t-conorm respectively.*

(1) *If T and S satisfy the distributivity:* $\forall x, y, z \in [0,1], S(x, T(y,z)) = T(S(x,y), S(x,z))$, *then* $T = T_{\min}$.

2.2 General Fuzzy Logic Connectives

(2) If T and S satisfy the distributivity: $\forall x, y, z \in [0,1]$, $T(x, S(y,z)) = S(T(x,y), T(x,z))$, then $S = S_{\max}$.

Proof. (1) Letting $z = 0$ yields $T(S(x,y), x) = x (\forall x, y \in [0,1])$. The desired result follows from Proposition 2.3(1). As for the proof of (2), let $z = 1$. □

Remark 2.4. *It can be seen from the proofs of Proposition 2.3 and Proposition 2.4 that the distributivity implies the absorption laws, and the absorption laws implies the idempotency for t−norms and t−conorms. The Zadeh's operations for forming intersection and union are the only choices if one of these properties (idempotency, absorption laws, distributivity) is required.*

Now we extend the union and intersection operations of fuzzy sets with the concepts of t−norm and t−conorm.

Definition 2.7. *Let T and S be a t-norm and a t-conorm respectively and n a strict negation. If $\forall x \in [0,1]$, $n(S(x,y)) = T(n(x), n(y))$, then (T, S, n) is called a De Morgan triple.*

For example, let $N(x) = 1 - x$. Then $(T_{\min}, S_{\max}, N)$, (T_L, S_L, N) and (T_π, S_π, N) are all De Morgan triples. It can be easily checked that both $(T_L^\varphi, S_L^\varphi, N_\varphi)$ and $(T_\pi^\varphi, S_\pi^\varphi, N_\varphi)$ are De Morgan triples which will be called a strong De Morgan triple and a strict De Morgan triple respectively.

Definition 2.8. *Let A, B be fuzzy sets on X and (T, S, n) a De Morgan triple. The complement A_n^c of A under n, the intersection $A \cap_T B$ of A and B under t−norm T and union $A \cup_S B$ of A and B under t−conorm S are respectively defined by:*

$$\forall x \in X, A_n^c(x) = n(A(x)),$$

$$\forall x \in X, (A \cap_T B)(x) = T(A(x), B(x))$$

and

$$(A \cup_S B)(x) = S(A(x), B(x)).$$

If $T = T_{\min}$, $n(x) = 1 - x$ and $S = S_{\max}$, then $(A \cap_T B)(x) = A(x) \wedge B(x)$ and $(A \cup_S B)(x) = A(x) \vee B(x)$, which are the Zadeh's intersection and union.

If $T = T_\pi$, $n(x) = 1 - x$ and $S = S_\pi$, then $(A \cap_T B)(x) = A(x)B(x)$ and $(A \cup_S B)(x) = A(x) + B(x) - A(x)B(x)$.

If $T = T_L$, $n(x) = 1 - x$ and $S = S_L$, then $(A \cap_T B)(x) = \max(0, A(x) + B(x) - 1)$ and $(A \cup_S B)(x) = \min(1, A(x) + B(x))$.

Proposition 2.5. *If (T, S, n) is a De Morgan triple, then the algebraic system $(F(X), \cup_S, \cap_T, c)$ has the following properties:*

(1) $A \cap_T B \subseteq A \subseteq A \cup_S B$;
(2) $A \cap_T B = B \cap_T A$, $A \cup_S B = B \cup_S A$;
(3) $(A \cap_T B) \cap_T C = A \cap_T (B \cap_T C)$, $(A \cup_S B) \cup_S C = A \cup_S (B \cup_S C)$;
(4) $A \cap_T \emptyset = \emptyset$, $A \cup_S \emptyset = A$, $A \cap_T X = A$, $A \cup_S X = X$;

(5) $(A \cup_S B)_n^c = A_n^c \cap_T B_n^c$. If n is a strong negation, then $(A \cap_T B)_n^c = A_n^c \cup_S B_n^c$.

Proof. All the proofs are straightforward. We show (5) as an example.

$$(A \cup_S B)_n^c(x) = n((A \cup_S B)(x)) = T(n(A(x)), n(B(x)))$$
$$= T(A_n^c(x), B_n^c(x)) = (A_n^c \cap_T B_n^c)(x).$$

If n is a strong negation,

$$(A \cap_T B)_n^c(x) = n((A \cap_T B)(x)) = n(T(A(x), B(x)))$$
$$= n(n(S(A_n^c(x), B_n^c(x)))) = (A_n^c \cup_S B_n^c)(x). \qquad \square$$

Remark 2.5. *There exist a lot of research on some special t-norms and t-conorms such as continuous, nilpotent or strict Archimedean ones [56, 71, 80, 81, 88, 110, 130]. Of all results in these studies, the most important are their representation theorems. For the mathematical representations of continuous Archimedean t-norms and t-conorms, see [88]. For the representations of nilpotent and strict Archimedean t-norms (t-conorms), see [110] and [130] respectively. In addition, t-norms and t-conorms can be unified in the framework of uninorms (see Exercise 24 for the definition of a uninorm). Concerning the study of uninorms, see, e.g. [31, 55, 81, 162].*

2.2.3 Fuzzy Implications

Definition 2.9. *Let $I : [0,1] \times [0,1] \to [0,1]$. If $I(x,y)$ is decreasing in x and increasing in y (usually I is called hybrid monotonous) and satisfies that $I(1,0) = 0$, $I(0,0) = I(1,1) = 1$, then I is called a fuzzy implication.*

It follows immediately from the definition that $I(0,1) = 1$, and thus a fuzzy implication is an extension of the ordinary implication in classic logic. The following are some examples of a fuzzy implication:

(1) $I_1(x,y) = \max(1-x, y)$,
(2) $I_2(x,y) = \min(1-x+y, 1)$,
(3) $I_3(x,y) = \begin{cases} 1 & x \leq y \\ \frac{y}{x} & x > y \end{cases}$.

It can be easily checked that if I is a fuzzy implication and n is a negation, then I' defined by $\forall x, y \in [0,1]$, $I'(x,y) = I(n(y), n(x))$ is a fuzzy implication as well.

Proposition 2.6. *A mapping $I : [0,1] \times [0,1] \to [0,1]$ is a fuzzy implication iff it satisfies the following:*

(I1) $\forall x \leq z, I(x,y) \geq I(z,y)$;
(I2) $\forall y \leq z, I(x,y) \leq I(x,z)$;
(I3) $\forall x \in [0,1], I(0,x) = 1$;

2.2 General Fuzzy Logic Connectives

(I4) $\forall x \in [0,1]$, $I(x,1) = 1$;
(I5) $I(1,0) = 0$.

Proof. The proof is straightforward. □

In the sequel, we introduce two special types of fuzzy implications, S-implications and R-implications.

Definition 2.10. *Let S be a t-conorm and N a strong negation. Then I defined by $\forall x,y \in [0,1]$, $I(x,y) = S(N(x), y)$ is called an S-implication.*

Remark 2.6. *In our definition of S-implication, N is required to be a strong negation. In [81], this requirement is dropped. In [4], the continuity of S is added and the corresponding implication is named as (S,N)-implication while the use of this name suggests a general logic connective I defined by: $\forall x,y \in [0,1]$, $I(x,y) = S(N(x), y)$ with T a t-norm and N a negation in [8, 9].*

It can be easily checked that $I(x,y) = S(N(x), y)$ is indeed a fuzzy implication. Here are some examples of an S-implication.

Kleene-Dienes implication:

$$S = S_{\max}, \quad N(x) = 1-x, \quad I(x,y) = \max(1-x, y).$$

Reichenbach implication:

$$S = S_\pi, \quad N(x) = 1-x, \quad I(x,y) = 1-x+xy.$$

Lukasiewicz implication:

$$S = S_L, \quad N(x) = 1-x, \quad I(x,y) = \min(1-x+y, 1).$$

Theorem 2.5. *An implication I is an S-implication iff I satisfies the following properties:*

(1) $\forall x \in [0,1]$, $I(1,x) = x$ (the so-called neutrality principle);
(2) $\forall x, y, z \in [0,1]$, $I(x, I(y,z)) = I(y, I(x,z))$ (the so-called exchange principle);
(3) There exists a strong negation N such that $\forall x, y \in [0,1]$, $I(x,y) = I(N(y), N(x))$.

Proof. Necessity. If $I(x,y) = S(N(x), y))$, where S is t-conorm and N a strong negation, then $I(1,x) = S(0,x) = x$, and thus (1) is valid.

In addition,

$$\begin{aligned} I(x, I(y,z)) &= S(N(x), I(y,z)) = S(N(x), S(N(y), z)) \\ &= S(S(N(y), z), N(x)) = S(N(y), S(z, N(x))) \\ &= I(y, I(x,z)). \end{aligned}$$

Hence (2) is true.

Finally, $I(N(y), N(x)) = S(y, N(x)) = I(x, y)$, i.e. (3) is true.

Sufficiency. Suppose that I satisfies (1), (2) and (3). Let

$$S(x, y) = I(N(x), y).$$

We prove that S is a t-conorm. Firstly, $S(0, y) = I(1, y) = y$, i.e. the boundary condition is satisfied. Since $S(x, y) = I(N(x), y) = I(N(y), x) = S(y, x)$ by (3), S is symmetric. Meanwhile, $\forall x, y, z \in [0, 1]$,

$$\begin{aligned} S(x, S(y, z)) &= I(N(x), S(y, z)) \\ &= I(N(x), I(N(y), z)) \\ &= I(N(x), I(N(z), y)) && \text{(by (3))} \\ &= I(N(z), I(N(x), y)) && \text{(by (2))} \\ &= I(N(I(N(x), y)), z) && \text{(by (3))} \\ &= I(N(S(x, y)), z) \\ &= S(S(x, y), z). \end{aligned}$$

Hence S is associative. In summary, S is a t-conorm. Noticing that $I(x, y) = S(N(x), y)$, I is an S-implication. $\square$

Definition 2.11. *Let T be a t-norm. Then I_T defined by: $\forall x, y \in [0, 1]$, $I_T(x, y) = \sup\{z | T(x, z) \leq y\}$ is called an R-implication.*

The definition is based on the following equality of crisp sets

$$A^c \cup B = (A \setminus B)^c = \cup \{X | A \cap X \subseteq B\}.$$

It should be pointed out that I_T is indeed a fuzzy implication.

Remark 2.7. *The R-implication is short for the residual implication, and I_T is also called the residuum of T (particularly for a left continuous T) in the literature.*

Example 2.6. *Let $T = T_{\min}$. Then we have the Gödel implication:*

$$I_T(x, y) = \begin{cases} 1 & x \leq y \\ y & x > y \end{cases}.$$

Let $T = T_\pi$. Then we have the Goguen implication:

$$I_T(x, y) = \begin{cases} 1 & x \leq y \\ \frac{y}{x} & x > y \end{cases}.$$

Let $T = T_L$. Then we have the Lukasiewicz implication:

$$I_T(x, y) = \min(1 - x + y, 1).$$

2.2 General Fuzzy Logic Connectives

Next we present a necessary and sufficient condition of R-implication based on a left continuous t-norm.

Lemma 2.3. *A t-norm T is left continuous iff* $(\forall x, y, z \in [0,1])$, $(T(x,z) \leq y \Leftrightarrow I_T(x,y) \geq z)$.

Proof. Necessity. If $T(x,z) \leq y$, it is clear that $I_T(x,y) \geq z$ by the definition of I_T.

If $I_T(x,y) \geq z$, we prove that $T(x,z) \leq y$.

When $I_T(x,y) > z$, there exists a $z_0 \in [0,1]$ such that $T(x,z_0) \leq y$ and $z_0 > z$. Hence $T(x,z) \leq T(x,z_0) \leq y$.

When $I_T(x,y) = z$, there exists a z_n such that $z \geq z_n > z - 1/n$ and $T(x,z_n) \leq y$. Taking the limit as $n \to \infty$ leads to $z_n \to z$. Noticing that $z_n < z$, we have $T(x,z) \leq y$ by the left-continuity of T.

Sufficiency. It suffices to prove that $\lim_{z' \uparrow z} T(x,z') = T(x,z)$ (by $z' \uparrow z$, we mean z' tends to z increasingly). Since $T(x,\cdot)$ is increasing, it is easily checked that $\lim_{z' \uparrow z} T(x,z') = \sup_{z' < z} T(x,z')$. As a result, we prove that $\sup_{z' < z} T(x,z') = T(x,z)$.

Let $y = \sup_{z' < z} T(x,z')$. When $z' < z$, $T(x,z') \leq y$, and hence

$$\sup_{z} \{z | T(x,z) \leq y\} \geq \sup_{z'} \{z | z' < z\} = z,$$

i.e. $I_T(x,y) \geq z$. According to our assumption,

$$T(x,z) \leq y = \sup_{z' < z} T(x,z').$$

Clearly, $T(x,z) \geq y$. Therefore,

$$T(x,z) = \sup_{z' < z} T(x,z'). \qquad \square$$

Lemma 2.4. *If the t-norm T is left continuous, then $I_T(x,y)$ is right continuous in y.*

Proof. We prove that $\lim_{y \downarrow y_0} I_T(x,y) = T(x,y_0)$ is valid for every $y_0 \in [0,1]$. Since $I_T(x,\cdot)$ is increasing, it suffices to show that $\inf_{y > y_0} I_T(x,y) = I_T(x,y_0)$. If the equality is violated, noticing that $\inf_{y > y_0} I_T(x,y) \geq I_T(x,y_0)$, then there exists an $\alpha \in [0,1]$ such that $\inf_{y > y_0} I_T(x,y) > \alpha > I_T(x,y_0)$. Hence $\forall y > y_0, I_T(x,y) > \alpha$. By the left-continuity of T and Lemma 2.3, we have $T(x,\alpha) \leq y (\forall y > y_0)$. By letting $y \to y_0$, it follows that $T(x,\alpha) \leq y_0$, whence $I_T(x,y_0) \geq \alpha$, a contradiction. $\qquad \square$

Theorem 2.6. $I : [0,1] \times [0,1] \to [0,1]$ *is an R-implication based on a left continuous t-norm iff I satisfies the following:*

(1) $I(x,y)$ is increasing in y;
(2) $I(x, I(y,z)) = I(y, I(x,z))$;
(3) $x \leq y \Leftrightarrow I(x,y) = 1$;
(4) $I(x,y)$ is right continuous in y.

Proof. Necessity. Suppose that $I(x,y) = I_T(x,y) = \sup\{z | T(x,z) \leq y\}$, where T is a left continuous t-norm. Clearly, $I(x,y)$ is increasing in y. By Lemma 2.4, $I(x,y)$ is right continuous in y. In addition, $I(x,y) = 1$ whenever $x \leq y$. Conversely, suppose that $I(x,y) = 1$. By the left continuity of T and Lemma 2.3, $T(x,1) \leq y$, i.e. $x \leq y$. Hence, $x \leq y \Leftrightarrow I(x,y) = 1$. What is left is the proof of (2). Indeed,

$$I(x, I(y,z)) = \sup\{t | T(x,t) \leq I(y,z)\}$$
$$= \sup\{t | T(y, T(x,t)) \leq z\} \quad \text{(by Lemma 2.3)}$$
$$= \sup\{t | T(T(x,y),t) \leq z\}$$
$$= I(T(x,y), z).$$

Therefore, $I(y, I(x,z)) = I(T(y,x),z) = I(T(x,y),z) = I(x, I(y,z))$, i.e. $I(x, I(y,z)) = I(y, I(x,z))$.

Sufficiency. Suppose that (1)-(4) are valid. We firstly prove that I is a fuzzy implication. Indeed, when $x \leq z$, $I(z, I(I(z,y),y)) = I(I(z,y), I(z,y)) = 1$ by (2) and (3), whence $z \leq I(I(z,y),y)$ by (3). Therefore,

$$I(I(z,y), I(x,y)) = I(x, I(I(z,y),y)) \geq I(x,z) = 1.$$

It follows from (3) that $I(z,y) \leq I(x,y)$. Hence, I satisfies the hybrid monotonicity due to (1). $I(0,0) = I(1,1) = 1$ follows immediately from (3). In addition,

$$I(1, I(I(1,0),0)) = I(I(1,0), I(1,0)) = 1.$$

By (3), $I(I(1,0), 0) = 1$, and thus $I(1,0) = 0$.

Secondly, let $T(x,y) = \inf\{t | I(x,t) \geq y\}$. We prove that T is a t-norm. Obviously, T is increasing by the hybrid monotonicity of I. Since $I(x, I(1,x)) = I(1, I(x,x)) = I(1,1) = 1$, $x \leq I(1,x)$. On the other hand,

$$I(1, I(I(1,x), x)) = I(I(1,x), I(1,x)) = 1,$$

whence $I(I(1,x), x) = 1$. It follows that $I(1,x) \leq x$ and thus $I(1,x) = x$. Therefore,

$$T(1,x) = \inf\{t | I(1,t) \geq x\} = \inf\{t | t \geq x\} = x,$$

i.e. the boundary condition is satisfied. By (3), we have successively

$$I(x,t) \geq y \Leftrightarrow I(y, I(x,t)) = 1 \Leftrightarrow I(x, I(y,t)) = 1 \Leftrightarrow I(y,t) \geq x.$$

2.2 General Fuzzy Logic Connectives

Hence,

$$T(x,y) = \inf\{t|I(x,t) \geq y\} = \inf\{t|I(y,t) \geq x\} = T(y,x),$$

which implies the symmetry of T. What is left is the proof of associativity of T. In order to verify $T(x,T(y,z)) = T(T(x,y),z) = T(z,T(x,y))$, it suffices to show that

$$I(x,t) \geq T(y,z) = T(z,y) \Leftrightarrow I(z,t) \geq T(x,y) = T(y,x).$$

Since $I(x,y) = I(x,y)$, we have $T(x,I(x,y)) \leq y$. From the right continuity of $I(\cdot,y)$, it can be shown that $I(x,T(x,y)) \geq y$ (Indeed, let $T(x,y) = \alpha$ and $\alpha_n = \alpha + \frac{1}{n}$. Then there exists t_n such that $I(x,t_n) \geq y$ and $\alpha \leq t_n < \alpha + \frac{1}{n}$. By letting $n \to \infty$, we have $I(x,\alpha) \geq y$ due to the right continuity of $I(\cdot,y)$).

Now it follows from $I(z,t) \geq T(x,y)$ that $I(x,I(z,t)) \geq I(x,T(x,y)) \geq y$. Hence,

$$T(z,y) \leq T(z,I(x,I(z,t))) = T(z,I(z,I(x,t))) \leq I(x,t).$$

Similarly, $I(z,t) \geq T(x,y)$ follows from $T(z,y) \geq I(x,t)$.

In summary, T is a t-norm.

Next, we prove that T is left continuous. It suffices to show that

$$\lim_{y \uparrow y_0} T(x,y) = T(x,y_0),$$

or equivalently

$$\bigvee_{y<y_0} T(x,y) = T(x,y_0) = \inf\{t|I(x,t) \geq y_0\},$$

since $T(x,\cdot)$ is increasing. If the equality is not valid, then there exists an α such that $\bigvee_{y<y_0} T(x,y) < \alpha < T(x,y_0)$. Hence, $\forall y < y_0$, $T(x,y) < \alpha < T(x,y_0)$, whence $y \leq I(x,T(x,y)) \leq I(x,\alpha)$. By letting $y \to y_0$, we have $I(x,\alpha) \geq y_0$, and thus $T(x,y_0) \leq \alpha$, a contradiction.

Finally, we prove that $I(x,y) = I_T(x,y) = \sup\{z|T(x,z) \leq y\}$.

Observing that $T(x,I(x,y)) \leq y$, we have $I_T(x,y) \geq I(x,y)$ by the definition of I_T.

On the other hand, $I(x,T(x,z)) \geq z$ is true for all $z \in [0,1]$. Particularly, let $z = I_T(x,y)$

$$I_T(x,y) \leq I(x,T(x,I_T(x,y))) \leq I(x,y),$$

since T is left continuous. Hence, $I_T(x,y) = I(x,y)$. □

Remark 2.8. *Most results in this subsection are adopted from [56]. Theorems 2.5 and 2.6 characterize S-implications and R-implications with T a* left continuous t-norm respectively. For the characterization of more general (S, N)-implications where N is continuous or strict, the reader may refer to [8]. For the characterizations of Lukasiewicz-like and Goguen-like implications, see [7] and [135] respectively. For the characterizations of fuzzy implications satisfying $I(x, y) = I(x, I(x, y))$, the reader is referred to [133].

2.2.4 Fuzzy Equivalencies

Definition 2.12. A mapping $E : [0, 1] \times [0, 1] \to [0, 1]$ is called a *fuzzy equivalence* if it satisfies that

(E1) $\forall x, y \in [0, 1]$, $E(x, y) = E(y, x)$;
(E2) $E(0, 1) = E(1, 0) = 0$;
(E3) $\forall x \in [0, 1]$, $E(x, x) = 1$;
(E4) $E(x, y) \leq E(x', y')$ whenever $x \leq x' \leq y' \leq y$.

Example 2.7. Lukasiewicz equivalence: $E(x, y) = 1 - |x - y|$.

Gödel equivalence: $E(x, y) = \begin{cases} 1 & x = y \\ \min(x, y) & x \neq y \end{cases}$.

Goguen equivalence: $E(x, y) = \begin{cases} 1 & x = y = 0 \\ \dfrac{\min(x, y)}{\max(x, y)} & \text{otherwise} \end{cases}$.

Proposition 2.7. *A mapping $E : [0, 1] \times [0, 1] \to [0, 1]$ is a fuzzy equivalence iff there exists a fuzzy implication I such that $\forall x \in [0, 1], I(x, x) = 1$ and $E(x, y) = \min(I(x, y), I(y, x))$.*

Proof. The proof of sufficiency is straightforward. We only prove the necessity.

Let $I(x, y) = \begin{cases} 1 & x \leq y \\ E(x, y) & x > y \end{cases}$. We verify that I is a fuzzy implication.

Firstly, we show that $I(x, y) \geq I(z, y)$ whenever $x \leq z$.
If $x \leq y$, $I(x, y) = 1$ and the desired equality trivially holds.
If $x > y$, then $y < x \leq z$,

$$I(x, y) = E(x, y) \geq E(z, y) = I(z, y).$$

Hence $I(\cdot, y)$ is decreasing. That $I(x, \cdot)$ is increasing in y can be similarly verified. When $x \leq y$, $I(x, y) = 1$, and $I(y, x) = E(y, x) = E(x, y)$. Hence,

$$E(x, y) = \min(I(x, y), I(y, x)).$$

In the case of $x > y$, the proof is similar. □

Corollary 2.1. *A mapping $E : [0, 1] \times [0, 1] \to [0, 1]$ is a fuzzy equivalence iff there exists a fuzzy implication I such that $\forall x \in [0, 1], I(x, x) = 1$, and*

$$E(x, y) = I(\max(x, y), \min(x, y)).$$

Proof. The proof of sufficiency is straightforward.

Necessity. It is obvious considering that

$$I(\max(x,y), \min(x,y)) = \min(I(x,y), I(y,x)).$$ □

2.3 Decomposition of a Fuzzy Set

As we know, a fuzzy set is an extension of a crisp set. Meanwhile, there exist close links between them. This and the next section are devoted to revealing these links from different angles. We first introduce the concept of an α−cut, one of the most fundamental concepts in fuzzy set theory.

Definition 2.13. *Let A be a fuzzy set on X. For $\alpha \in [0,1]$, the α-cut A_α of A is defined as $A_\alpha = \{x|A(x) \geq \alpha\}$, and the strong α-cut $A_{\dot\alpha}$ of A is defined as $A_{\dot\alpha} = \{x|A(x) > \alpha\}$.*

Hence the (strong) α-cut of a fuzzy set A is a crisp subset of X with the membership degree of every element greater than or equal to (resp. greater than) α. For example, if a fuzzy set A on $X = \{x_1, x_2, x_3, x_4, x_5, x_6\}$ is defined as $A = 0.2/x_1 + 0.7/x_2 + 0.8/x_3 + 0.3/x_4 + 0.4/x_5 + 0.4/x_6$. Then $A_{0.4} = \{x_2, x_3, x_5, x_6\}$ and $A_{\dot{0.4}} = \{x_2, x_3\}$.

Example 2.8. *Let a fuzzy set A on $[a,d] (a,d \in \mathbb{R})$ be defined by*

$$A(x) = \begin{cases} \frac{x-a}{b-a} & a < x < b \\ 1 & b \leq x \leq c \\ \frac{x-d}{c-d} & c < x < d \end{cases},$$

where a, b, c, d are real numbers.
Then, for any $\alpha \in [0,1]$, $A_\alpha = [a+\alpha(b-a), d+\alpha(c-d)]$ and $A_{\dot\alpha} =]a+\alpha(b-a), d+\alpha(c-d)[$.

Theoretically, it is more reasonable to employ fuzzy sets to describe the performance or concepts of the involved objects in many real world problems. However, we have to make a final decision, e.g. choice, ranking, clustering and recognition, when confronted with fuzzy information. At this point, it is natural for us to prescribe a threshold α and not to take into account objects with the membership degrees lower than the threshold. α-cuts of a fuzzy set are thus often useful.

Example 2.9. *In the study of Chinese history, the slave society A may be appropriately defined as the fuzzy set on the set of dynasties: $A = 1/Xia + 1/Shang + 0.9/Zhou + 0.7/Chunqiou + 0.5/Zhanguo + 0.4/Qin + 0.3/Xihan + 0.1/Donghan$. If we have to make crisp classification, it may be reasonable* to take the threshold as 0.5. Then $A_{0.5} = \{Xia, Shang, Zhou, Chunqiou, Zhanguo\}$ is selected as a set of representative dynasties of the slave society.

From Definition 2.13, it follows immediately that $A_{\dot{\alpha}} \subseteq A_\alpha$ ($\forall \alpha \in [0,1]$), $A_0 = X$, $A_{\dot{1}} = \emptyset$, $A_1 = \ker(A)$ and $A_{\dot{0}} = \text{supp}(A)$. In the following, we further discuss properties of (strong) α-cuts.

Proposition 2.8. Let $A, B, A_i, B_i \in F(X) (i \in I)$. Then

(1) $(A \cup B)_\alpha = A_\alpha \cup B_\alpha \qquad (A \cap B)_\alpha = A_\alpha \cap B_\alpha$
$(A \cup B)_{\dot{\alpha}} = A_{\dot{\alpha}} \cup B_{\dot{\alpha}} \qquad (A \cap B)_{\dot{\alpha}} = A_{\dot{\alpha}} \cap B_{\dot{\alpha}}$

(2) $\bigcup\limits_{i \in I}(A_i)_\alpha \subseteq (\bigcup\limits_{i \in I} A_i)_\alpha \qquad \bigcap\limits_{i \in I}(A_i)_\alpha = (\bigcap\limits_{i \in I} A_i)_\alpha$
$(\bigcup\limits_{i \in I} A_i)_{\dot{\alpha}} = \bigcup\limits_{i \in I}(A_i)_{\dot{\alpha}} \qquad (\bigcap\limits_{i \in I} A_i)_{\dot{\alpha}} \subseteq \bigcap\limits_{i \in I}(A_i)_{\dot{\alpha}}$

(3) If $\alpha_1 < \alpha_2$, then $A_{\alpha_2} \subseteq A_{\alpha_1}$, $A_{\alpha_2} \subseteq A_{\dot{\alpha}_1}$ and $A_{\dot{\alpha}_2} \subseteq A_{\dot{\alpha}_1}$.

(4) Let $\alpha = \bigvee\limits_{i \in I} \alpha_i$, $\beta = \bigwedge\limits_{i \in I} \alpha_i$. Then $\bigcap\limits_{i \in I} A_{\alpha_i} = A_\alpha \qquad \bigcup\limits_{i \in I} A_{\dot{\alpha}_i} = A_{\dot{\beta}}$.
Particularly, $\bigcap\limits_{\lambda < \alpha} A_\lambda = A_\alpha$ and $\bigcup\limits_{\lambda > \alpha} A_\lambda = A_{\dot{\alpha}}$.

(5) $(A^c)_\alpha = (A_{\dot{1-\alpha}})^c \qquad (A^c)_{\dot{\alpha}} = (A_{1-\alpha})^c$.

Proof. We merely give the proofs of some equalities and inclusions. The others are left to the reader.

(1) $\forall x \in X$, $\forall \alpha \in [0,1]$,

$$\begin{aligned}
x \in (A \cup B)_\alpha &\Leftrightarrow (A \cup B)(x) \geq \alpha \\
&\Leftrightarrow \max(A(x), B(x)) \geq \alpha \\
&\Leftrightarrow A(x) \geq \alpha \text{ or } B(x) \geq \alpha \\
&\Leftrightarrow x \in A_\alpha \text{ or } x \in B_\alpha \\
&\Leftrightarrow x \in A_\alpha \cup B_\alpha,
\end{aligned}$$

which completes the proof of $(A \cup B)_\alpha = A_\alpha \cup B_\alpha$.

(2) $\forall x \in X$, $\forall \alpha \in [0,1]$,

$$\begin{aligned}
x \in \bigcup\limits_{i \in I}(A_i)_\alpha &\Rightarrow \exists i \in I, x \in (A_i)_\alpha \\
&\Rightarrow \exists i \in I, A_i(x) \geq \alpha \\
&\Rightarrow \sup\limits_{i \in I} A_i(x) \geq \alpha \\
&\Rightarrow x \in (\bigcup\limits_{i \in I} A_i)_\alpha.
\end{aligned}$$

Thus $\bigcup\limits_{i \in I}(A_i)_\alpha \subseteq (\bigcup\limits_{i \in I} A_i)_\alpha$.

(3) Obvious.

2.3 Decomposition of a Fuzzy Set

(4) $\forall x \in X$,

$$x \in \bigcap_{i \in I} A_{\alpha_i} \Leftrightarrow \forall i \in I, x \in A_{\alpha_i}$$

$$\Leftrightarrow \forall i \in I, A(x) \geq \alpha_i$$

$$\Leftrightarrow A(x) \geq \bigvee_{i \in I} \alpha_i = \alpha$$

$$\Leftrightarrow x \in A_\alpha.$$

Hence $\bigcap_{i \in I} A_{\alpha_i} = A_\alpha$.

(5) $x \in (A_{1-\alpha})^c \Leftrightarrow x \notin A_{1-\alpha} \Leftrightarrow A(x) \leq 1 - \alpha \Leftrightarrow A^c(x) \geq \alpha \Leftrightarrow x \in (A^c)_\alpha$.
Hence $(A^c)_\alpha = (A_{1-\alpha})^c$. □

It should be noted that either of the two inclusions in Proposition 2.8(2) can not be changed into equalities. As an illustration, we present an example in which the equality $\bigcup_{i \in I}(A_i)_\alpha = (\bigcup_{i \in I} A_i)_\alpha$ does not hold.

Example 2.10. Let A_n $(n = 1, 2, \ldots)$ be fuzzy sets on a universal set X, defined by

$$\forall x \in X, A_n(x) = 0.5 - \frac{1}{n+1}.$$

Put $\alpha = 0.5$. Then for every $x \in X$ and $n = 1, 2, \cdots$, $A_n(x) < \alpha$ and hence $(A_n)_\alpha = \emptyset$. As a consequence,

$$\bigcup_{n=1}^{\infty} (A_n)_\alpha = \emptyset.$$

However,

$$(\bigcup_{n=1}^{\infty} A_n)(x) = \sup_{n=1,2,\ldots} A_n(x) = 0.5$$

and hence $x \in (\bigcup_{n=1}^{\infty} A_n)_\alpha$ for every $x \in X$. Therefore,

$$(\bigcup_{n=1}^{\infty} A_n)_\alpha = X.$$

Clearly, $(\bigcup_{n=1}^{\infty} A_n)_\alpha \neq \bigcup_{n=1}^{\infty} (A_n)_\alpha$.

For $\alpha \in [0, 1]$ and $A \in F(X)$, $\alpha A \in F(X)$ is defined by

$$\forall x \in X, \quad (\alpha A)(x) = \alpha \wedge A(x).$$

This operation has two basic properties:

(i) $\alpha_1 < \alpha_2$ implies that $\alpha_1 A \subseteq \alpha_2 A$;
(ii) $A_1 \subseteq A_2$ implies that $\alpha A_1 \subseteq \alpha A_2$.

Even though A is a crisp set, αA is still a fuzzy set with the membership function defined by

$$(\alpha A)(x) = \begin{cases} \alpha & x \in A \\ 0 & \text{otherwise.} \end{cases}$$

Now we present three decomposition theorems.

Theorem 2.7. *(Decomposition Theorem I) For every $A \in F(X)$,*

$$A = \bigcup_{\alpha \in [0,1]} \alpha A_\alpha.$$

Proof. $\forall x \in X$,

$$\left(\bigcup_{\alpha \in [0,1]} \alpha A_\alpha\right)(x) = \bigvee_{\alpha \in [0,1]} (\alpha A_\alpha)(x) = \bigvee_{\alpha \in [0,1]} (\alpha \wedge A_\alpha(x))$$

$$= \max\left\{ \bigvee_{x \in A_\alpha} (\alpha \wedge A_\alpha(x)),\ \bigvee_{x \notin A_\alpha} (\alpha \wedge A_\alpha(x)) \right\}$$

$$= \bigvee_{\alpha \leq A(x)} \alpha = A(x). \qquad \square$$

Theorem 2.8. *(Decomposition Theorem II) For every $A \in F(X)$,*

$$A = \bigcup_{\alpha \in [0,1]} \alpha A_{\underline{\alpha}}.$$

Proof. Similar to the proof of Decomposition Theorem I. $\qquad \square$

Corollary 2.2. *Let A and B be fuzzy sets on X. Then $A = B \Leftrightarrow \forall \alpha \in [0,1]$, $A_\alpha = B_\alpha \Leftrightarrow \forall \alpha \in [0,1]$, $A_{\underline{\alpha}} = B_{\underline{\alpha}}$.*

Proof. Trivial. $\qquad \square$

By Corollary 2.2, a fuzzy set and all its (strong) α-cuts are uniquely determined by each other. As a matter of fact, it can be seen that $A(x) = \bigvee_{x \in A_\alpha} \alpha$ from the proof of Decomposition Theorem I. Similarly $A(x) = \bigvee_{x \in A_{\underline{\alpha}}} \alpha$ based on Decomposition Theorem II. So we can find the fuzzy set A if its (strong) α-cuts are given for all $\alpha \in [0,1]$.

Example 2.11. $X = \{x_1, x_2, x_3, x_4, x_5\}$.

$$A_\alpha = \begin{cases} X & 0 \leq \alpha \leq 0.2 \\ \{x_1, x_2, x_3, x_5\} & 0.2 < \alpha \leq 0.5 \\ \{x_1, x_3, x_5\} & 0.5 < \alpha \leq 0.6 \\ \{x_1, x_3\} & 0.6 < \alpha \leq 0.7 \\ \{x_3\} & 0.7 < \alpha \leq 1 \end{cases}$$

Applying $A(x) = \bigvee_{x \in A_\alpha} \alpha$, we have

$$A(x_1) = 0.7, A(x_2) = 0.5, A(x_3) = 1, A(x_4) = 0.2, A(x_5) = 0.6.$$

The resulting fuzzy set is $A = 0.7/x_1 + 0.5/x_2 + 1/x_3 + 0.2/x_4 + 0.6/x_5$.

For convenience, we make the following conventions henceforth:

For a family of sets $A_i \in P(X)(i \in I)$, $\bigcap_{i \in \emptyset} A_i = X$ and $\bigcup_{i \in \emptyset} A_i = \emptyset$.

Theorem 2.9. *(Decomposition Theorem III) If $H : [0,1] \to P(X)$ satisfies that $\forall \alpha \in [0,1]$, $A_{\dot\alpha} \subseteq H(\alpha) \subseteq A_\alpha$, then $A = \bigcup_{\alpha \in [0,1]} \alpha H(\alpha)$ and the following properties are fulfilled:*

(1) $H(\alpha_1) \supseteq H(\alpha_2)$ whenever $\alpha_1 < \alpha_2$.
(2) $\forall \alpha \in [0,1], A_\alpha = \bigcap_{\lambda < \alpha} H(\lambda)$.
(3) $\forall \alpha \in [0,1], A_{\dot\alpha} = \bigcup_{\lambda > \alpha} H(\lambda)$.

Proof. $A = \bigcup_{\alpha \in [0,1]} \alpha A_{\dot\alpha} \subseteq \bigcup_{\alpha \in [0,1]} \alpha H(\alpha) \subseteq \bigcup_{\alpha \in [0,1]} \alpha A_\alpha = A$ due to Decomposition Theorems I and II. Hence $A = \bigcup_{\alpha \in [0,1]} \alpha H(\alpha)$.

Now we turn to the proof of (1), (2) and (3).

(1) Assume $\alpha_1 < \alpha_2$. Then

$$H(\alpha_2) \subseteq A_{\alpha_2} \subseteq A_{\dot\alpha_1} \subseteq H(\alpha_1).$$

(2) If $\lambda < \alpha$, then $H(\lambda) \supseteq A_\lambda \supseteq A_\alpha$, hence $A_\alpha \subseteq \bigcap_{\lambda < \alpha} H(\lambda)$. On the other hand, $\bigcap_{\lambda < \alpha} H(\lambda) \subseteq \bigcap_{\lambda < \alpha} A_\lambda = A_\alpha$ by Proposition 2.8(4). Therefore, $A_\alpha = \bigcap_{\lambda < \alpha} H(\lambda)$.

(3) The reasoning is similar to that in the proof of (2). □

2.4 Mathematical Representation of Fuzzy Sets

In this section, we establish the links between a fuzzy set and a nest of sets whose definition is given as follows.

Definition 2.14. *If a mapping H from $[0,1]$ to $P(X)$ satisfies that $H(\alpha_1) \supseteq H(\alpha_2)$ whenever $\alpha_1 < \alpha_2$, then H is called a nest of sets on X. The set of all nests of sets on X will be denoted by $U(X)$.*

For $\alpha \in [0,1]$, define $H_1(\alpha) = A_\alpha$, $H_2(\alpha) = A_{\dot{\alpha}}$. Then H_1 and H_2 are two nests of sets by Proposition 2.8(3). In addition, $\dot{H}$ in Decomposition Theorem III is a nest of sets.

Let H, H_1, H_2 be three nests of sets. The operations $\cup, \cap$ and c in $U(X)$ are defined by

$\forall \alpha \in [0,1]$, $(H_1 \cup H_2)(\alpha) = H_1(\alpha) \cup H_2(\alpha)$;
$\forall \alpha \in [0,1]$, $(H_1 \cap H_2)(\alpha) = H_1(\alpha) \cap H_2(\alpha)$;
$\forall \alpha \in [0,1]$, $H^c(\alpha) = (H(1-\alpha))^c$.

It is easily verified that $H_1 \cup H_2$, $H_1 \cap H_2$ and H^c are indeed nests of sets.

Proposition 2.9. *$(U(X), \cup, \cap, c)$ is a soft algebra.*

Proof. We prove the De Morgan law: $(H_1 \cap H_2)^c = H_1^c \cup H_2^c$ as an example. For every $\alpha \in [0,1]$,

$$(H_1 \cap H_2)^c(\alpha) = ((H_1 \cap H_2)(1-\alpha))^c = (H_1(1-\alpha) \cap H_2(1-\alpha))^c$$
$$= (H_1(1-\alpha))^c \cup (H_2(1-\alpha))^c = H_1^c(\alpha) \cup H_2^c(\alpha)$$
$$= (H_1^c \cup H_2^c)(\alpha),$$

i.e. $(H_1 \cap H_2)^c = H_1^c \cup H_2^c$. □

In the soft algebra $(U(X), \cup, \cap, c)$, the greatest element $\overline{X}$ and the least element $\overline{\emptyset}$ are respectively defined by

$$\forall \alpha \in [0,1], \overline{X}(\alpha) = X \quad \text{and} \quad \overline{\emptyset}(\alpha) = \emptyset.$$

Theorem 2.10. *(Representation Theorem) If a mapping f from $U(X)$ to $F(X)$ is defined by $\forall H \in U(X)$, $f(H) = \bigcup_{\alpha \in [0,1]} \alpha H(\alpha)$, then f is a surjective homomorphism from $(U(X), \cup, \cap, c)$ to $(F(X), \cup, \cap, c)$ such that*

(1) $\forall \lambda \in [0,1]$, $(f(H))_{\dot{\lambda}} \subseteq H(\lambda) \subseteq (f(H))_\lambda$;
(2) $\forall \lambda \in [0,1]$, $(f(H))_\lambda = \bigcap_{\alpha < \lambda} H(\alpha)$;
(3) $\forall \lambda \in [0,1]$, $(f(H))_{\dot{\lambda}} = \bigcup_{\alpha > \lambda} H(\alpha)$.

Proof. Firstly we show that f is surjective.

For every $A \in F(X)$, let $H(\alpha) = A_\alpha (\forall \alpha \in [0,1])$. Then $H \in U(X)$. By Decomposition Theorem I,

$$f(H) = \bigcup_{\alpha \in [0,1]} \alpha H(\alpha) = \bigcup_{\alpha \in [0,1]} \alpha A_\alpha = A,$$

which indicates that f is a surjection.

2.4 Mathematical Representation of Fuzzy Sets

Next we prove that the inclusions in (1) are valid.
$\forall x \in H(\lambda)$, $(f(H))(x) = \bigvee_{\alpha \in [0,1]} \alpha \wedge (H(\alpha))(x) \geq \lambda \wedge (H(\lambda))(x) = \lambda$. Hence $x \in (f(H))_\lambda$, and thus $H(\lambda) \subseteq (f(H))_\lambda$.

If $x \notin H(\lambda)$, then $x \notin H(\alpha)$ holds for all $\alpha > \lambda$ since H is a nest of set. Therefore,

$$(f(H))(x) = \bigvee_{\alpha \in [0,1]} \alpha \wedge (H(\alpha))(x) = \bigvee_{\alpha < \lambda} \alpha \wedge (H(\alpha))(x) \leq \bigvee_{\alpha < \lambda} \alpha = \lambda,$$

which means $x \notin (f(H))_{\dot\lambda}$. Consequently, $(f(H))_{\dot\lambda} \subseteq H(\lambda)$. Now (2) and (3) follow from Decomposition Theorem III.

Finally we verify that f is operations-preserving.
By (3), $\forall \lambda \in [0,1]$,

$$(f(H_1 \cup H_2))_{\dot\lambda} = \bigcup_{\alpha > \lambda}(H_1 \cup H_2)(\alpha) = (\bigcup_{\alpha > \lambda} H_1(\alpha)) \cup (\bigcup_{\alpha > \lambda} H_2(\alpha))$$
$$= (f(H_1))_{\dot\lambda} \cup (f(H_2))_{\dot\lambda} = (f(H_1) \cup f(H_2))_{\dot\lambda}.$$

Therefore, $f(H_1 \cup H_2) = f(H_1) \cup f(H_2)$ by Corollary 2.2. In other words, f is union-preserving.

Similarly, $\forall \lambda \in [0,1]$,

$$(f(H_1 \cap H_2))_\lambda = \bigcap_{\alpha < \lambda}(H_1 \cap H_2)(\alpha) = (\bigcap_{\alpha < \lambda} H_1(\alpha)) \cap (\bigcap_{\alpha < \lambda} H_2(\alpha))$$
$$= (f(H_1))_\lambda \cap (f(H_2))_\lambda = (f(H_1) \cap f(H_2))_\lambda$$

and

$$(f(H^c))_\lambda = \bigcap_{\alpha < \lambda} H^c(\alpha) = \bigcap_{\alpha < \lambda}(H(1-\alpha))^c$$
$$= (\bigcup_{\alpha < \lambda} H(1-\alpha))^c = (\bigcup_{\alpha' > 1-\lambda} H(\alpha'))^c$$
$$= [(f(H))_{1-\lambda}]^c = [(f(H))^c]_\lambda.$$

Hence, $f(H_1 \cap H_2) = f(H_1) \cap f(H_2)$ and $(f(H))^c = f(H^c)$, i.e. f also preserves intersection and complement. □

Using the surjective homomorphism f, we define the relation $\sim$ on $U(X)$ by

$$H_1 \sim H_2 \Leftrightarrow f(H_1) = f(H_2).$$

Then $\sim$ is an equivalence relation and the corresponding equivalence class with representant H is $[H] = \{H' | H' \sim H\} = \{H' | f(H') = f(H)\}$ and the quotient set $U(X)/\sim = \{[H]|H \in U(X)\}$. Meanwhile, formulate the operations $\cup, \cap, c$ in $U(X)/\sim$ as follows:

$\forall H_1, H_2 \in U(X), [H_1] \cup [H_2] = [H_1 \cup H_2]$;
$\forall H_1, H_2 \in U(X), [H_1] \cap [H_2] = [H_1 \cap H_2]$;
$\forall H \in U(X), [H]^c = [H^c]$.

An induced mapping f' from $U(X)/\sim$ to $F(X)$ from f is defined by

$$\forall [H] \in U(X)/\sim, \ f'([H]) = f(H).$$

By Proposition 1.5, f' is an isomorphism between $(U(X)/\sim, \cup, \cap, c)$ and $(F(X), \cup, \cap, c)$. In this sense, a fuzzy set may be viewed as an equivalence class of a nest of sets.

2.5 L-Fuzzy Sets

In the definition of a fuzzy set, the range of the involved mapping is confined to the totally ordered set $[0, 1]$. From the mathematical view, this restriction is not natural. In this section, $[0, 1]$ is extended to a general lattice L, which leads to the so-called L-fuzzy sets. As in fuzzy sets, some operations such as union and intersection may be formed by employing the concept of supremum and infimum in L. However, a generalization of the complement operation needs some extra efforts since there is no operation in L available for formulating complement. In view of this, we start with the concept of a pseudo-complement.

2.5.1 Pseudo-complements

Definition 2.15. Let $(P, \leq)$ be a poset. If a mapping n from P to P satisfies that

(1) $\forall \alpha, \beta \in P$, $\alpha \leq \beta$ implies that $n(\beta) \leq n(\alpha)$,
(2) $\forall \alpha \in P$, $n(n(\alpha)) = \alpha$,

then n is called a pseudo-complement on $(P, \leq)$.

Clearly, every strong negation is a pseudo-complement on $([0, 1], \leq)$ and $n(A) = A^c$ ($\forall A \in P(X)$) is a pseudo-complement on $(P(X), \subseteq)$.

Example 2.12. The complement c in a soft algebra $(L, \vee, \wedge, c)$ is a pseudo-complement. Since the complement c in every soft algebra is involutive, it suffices to prove that $\forall \alpha, \beta \in P$, $\alpha \leq \beta$ implies that $\beta^c \leq \alpha^c$. Indeed, when $\alpha \leq \beta$, $\alpha^c \vee \beta^c = (\alpha \wedge \beta)^c = \alpha^c$ by De Morgan laws. Hence $\beta^c \leq \alpha^c$.

The complement in a Boolean algebra is also a pseudo-complement since every Boolean algebra is a soft algebra.

2.5 L-Fuzzy Sets

Proposition 2.10. *If $(P, \leq)$ is a bounded poset with the greatest element 1 and the least element 0 and if n is a pseudo-complement on $(P, \leq)$, then $n(1) = 0$ and $n(0) = 1$.*

Proof. From $n(0) \leq 1$, it follows that $n(n(0)) \geq n(1)$, namely $n(1) \leq 0$. Hence $n(1) = 0$. The identity $n(0) = 1$ can be similarly verified. □

Proposition 2.11. *If n is a pseudo-complement in a lattice $(L, \vee, \wedge)$, then*

$$n(\alpha \vee \beta) = n(\alpha) \wedge n(\beta)$$

and

$$n(\alpha \wedge \beta) = n(\alpha) \vee n(\beta).$$

Proof. It follows from $\alpha \leq \alpha \vee \beta$ and $\beta \leq \alpha \vee \beta$ that $n(\alpha) \geq n(\alpha \vee \beta)$ and $n(\beta) \geq n(\alpha \vee \beta)$. Thus

$$n(\alpha \vee \beta) \leq n(\alpha) \wedge n(\beta).$$

Similarly, $n(\alpha \wedge \beta) \geq n(\alpha) \vee n(\beta)$, which implies that

$$n(n(\alpha) \wedge n(\beta)) \geq n(n(\alpha)) \vee n(n(\beta)) = \alpha \vee \beta,$$

i.e.

$$n(\alpha \vee \beta) \geq n(\alpha) \wedge n(\beta).$$

By antisymmetry,

$$n(\alpha \vee \beta) = n(\alpha) \wedge n(\beta).$$

The other equality can be verified following a similar reasoning.

For a complete lattice, the preceding proposition can be extended as:

$$n(\bigvee_{i \in I} \alpha_i) = \bigwedge_{i \in I} n(\alpha_i)$$

and

$$n(\bigwedge_{i \in I} \alpha_i) = \bigvee_{i \in I} n(\alpha_i),$$

where I is an arbitrary index set.

2.5.2 L-Fuzzy Sets and Their Set-Theoretic Operations

Definition 2.16. *Let X be the universe of discourse and let $(L, \vee, \wedge)$ be a lattice. A mapping A from X to L is said to be an L-fuzzy set on X. The set of all L-fuzzy sets on X will be denoted by $F_L(X)$.*

The concept of an L-fuzzy set was firstly introduced by Goguen[62]. He pointed out that $F_L(X)$ can be given whatever operations L has, and these operations will obey any law valid in L which extends point by point. For example, the concepts of subset, union and intersection can be defined by means of $\leq$, $\vee$ and $\wedge$ in L respectively. More specifically, let $A, B \in F_L(X)$.

If $\forall x \in X$, $A(x) \leq B(x)$, then A is called a subset of B, denoted by $A \subseteq B$. The union $A \cup B$ of A and B is defined by

$$\forall x \in X, (A \cup B)(x) = A(x) \vee B(x).$$

The intersection $A \cap B$ of A and B is defined by

$$\forall x \in X, (A \cap B)(x) = A(x) \wedge B(x).$$

Clearly, $A = B$ iff $A \subseteq B$ and $B \subseteq A$.

For a complete lattice $(L, \vee, \wedge)$ and $A_i \in F_L(X)$ $(i \in I)$, union and intersection can be extended,

$$\forall x \in X, (\bigcup_{i \in I} A_i)(x) = \bigvee_{i \in I} A_i(x) \quad \text{and} \quad (\bigcap_{i \in I} A_i)(x) = \bigwedge_{i \in I} A_i(x)$$

If there is a pseudo-complement n on $(L, \leq)$, then the complement A^c of A in $F_L(X)$ is defined by

$$\forall x \in X, A^c(x) = n(A(x)).$$

Generally speaking, $(F_L(X), \cup, \cap)$ is a lattice. As some additional conditions are imposed on L, $F_L(X)$ will gain some more properties. As examples, we list the following:

(1) If $(L, \vee, \wedge)$ is a distributive lattice, then $(F_L(X), \cup, \cap)$ is a distributive lattice.
(2) If $(L, \vee, \wedge)$ is a bounded distributive lattice and there exists a pseudo-complement n on $(L, \leq)$, then $(F_L(X), \cup, \cap, c)$ is a soft algebra.
(3) If $(L, \vee, \wedge, n)$ is a Boolean algebra, then $(F_L(X), \cup, \cap, c)$ is a Boolean algebra.
(4) If $(L, \vee, \wedge, n)$ is a superior soft algebra, then $(F_L(X), \cup, \cap, c)$ is a superior soft algebra.

We leave the proof of these assertions to the reader.

Example 2.13. Let $L = P(Y)$ and $\vee, \wedge$ and n be the union, intersection and complement of crisp sets respectively. Then $(F_L(X), \cup, \cap, c)$ is a Boolean algebra.

Example 2.14. Let $L = \{[a, b] | a \leq b, a, b \in [0, 1]\}$. Define a relation $\leq$ and a pseudo-complement n respectively by

$$[a_1, b_1] \leq [a_2, b_2] \Leftrightarrow a_1 \leq a_2 \text{ and } b_1 \leq b_2$$

2.5 L-Fuzzy Sets

and
$$n([a,b]) = [1-b, 1-a].$$

It can be easily checked that $(L, \vee, \wedge, n)$ is a superior soft algebra. Hence $(F_L(X), \cup, \cap, c)$ is also a superior soft algebra.

Example 2.15. Let $L = F([0,1])$ and $\vee, \wedge$ and n be the union, intersection and complement of fuzzy sets respectively. Then $(F_L(X), \cup, \cap, c)$ is a superior soft algebra. $A \in F_L(X)$ is usually called a type-2 fuzzy set. In a recursive manner, the concept of a fuzzy set of higher type may be introduced.

2.5.3 Decomposition of an L-Fuzzy Set

For $A \in F_L(X)$ and $\alpha \in L$, $A_\alpha = \{x | \alpha \le A(x)\}$ and $A_{\dot\alpha} = \{x | \alpha < A(x)\}$ will be called the α-cut and strong α-cut of A respectively. It is apparent that (1) $A_\alpha \supseteq A_{\dot\alpha}$ (2) $\alpha < \beta$ implies that $A_\alpha \supseteq A_\beta$, $A_{\dot\alpha} \supseteq A_\beta$ and $A_{\dot\alpha} \supseteq A_{\dot\beta}$. In addition, we have some more properties.

Proposition 2.12. If L is a complete lattice and $A_i \in F_L(X)$ $(i \in I)$, then

(1) $\bigcup_{i \in I}(A_i)_\alpha \subseteq (\bigcup_{i \in I} A_i)_\alpha$;

(2) $\bigcap_{i \in I}(A_i)_\alpha = (\bigcap_{i \in I} A_i)_\alpha$;

(3) $(\bigcup_{i \in I} A_i)_{\dot\alpha} \supseteq \bigcup_{i \in I}(A_i)_{\dot\alpha}$;

(4) $(\bigcap_{i \in I} A_i)_{\dot\alpha} \subseteq \bigcap_{i \in I}(A_i)_{\dot\alpha}$.

Proof. All the proofs are similar to those of Proposition 2.8 and are omitted here.

It is worth noting that Proposition 2.12(3) is no more an equality as in the case of fuzzy set even though I is finite as can be seen in the following example.

Example 2.16. Let $X = \{x_1, x_2, x_3\}$ and $(L, \vee, \wedge) = (P(X), \cup, \cap)$. Suppose that $A = \{x_1\}$, $B = \{x_2, x_3\}$, $C = \{x_1, x_2\}$ and $A_1, A_2 \in F_L(X)$ are defined by: $\forall x \in X$, $A_1(x) = A$ and $A_2(x) = B$ respectively. Set $\alpha = C$. Then

$$(A_1)_{\dot\alpha} = \{x | A_1(x) > \alpha\} = \{x | A_1(x) \supset C\} = \emptyset.$$

Similarly, $(A_2)_{\dot\alpha} = \emptyset$ and thus $(A_1)_{\dot\alpha} \cup (A_2)_{\dot\alpha} = \emptyset$. However,

$$\forall x \in X, (A_1 \cup A_2)(x) = A_1(x) \cup A_2(x) = A \cup B = X.$$

Hence, $(A_1 \cup A_2)_{\dot\alpha} = X$, and thus

$$(A_1 \cup A_2)_{\dot\alpha} \ne (A_1)_{\dot\alpha} \cup (A_2)_{\dot\alpha}.$$

Proposition 2.13. *Let L be a complete lattice. For $\alpha_i \in L$ ($i \in I$), write $\alpha = \bigvee_{i \in I} \alpha_i$ and $\beta = \bigwedge_{i \in I} \alpha_i$. Then*

(1) $A_\alpha = \bigcap_{i \in I} A_{\alpha_i}$,

(2) $A_\beta \supseteq \bigcup_{i \in I} A_{\alpha_i}$,

(3) $A_{\dot\alpha} \subseteq \bigcap_{i \in I} A_{\dot\alpha_i}$,

(4) $A_{\dot\beta} \supseteq \bigcup_{i \in I} A_{\dot\alpha_i}$.

Proof. As an example, we show (1).

$x \in A_\alpha \Leftrightarrow A(x) \geq \alpha = \bigvee_{i \in I} \alpha_i \Leftrightarrow \forall i \in I, A(x) \geq \alpha_i \Leftrightarrow \forall i \in I, x \in A_{\alpha_i} \Leftrightarrow x \in \bigcap_{i \in I} A_{\alpha_i}$.

Corollary 2.3. *For any $A \in F_L(X)$, $A_\alpha = \bigcap_{\lambda \leq \alpha} A_\lambda$ and $A_{\dot\alpha} = \bigcup_{\lambda \geq \alpha} A_\lambda$. If L is dense, then $A_\alpha = \bigcap_{\lambda < \alpha} A_\lambda$.*

Proof. $A_\alpha = \bigcap_{\lambda \leq \alpha} A_\lambda$ follows from Proposition 2.13(1). By Proposition 2.13(4), $A_{\dot\alpha} \supseteq \bigcup_{\lambda \geq \alpha} A_\lambda$. The reverse inclusion $A_{\dot\alpha} \subseteq \bigcup_{\lambda \geq \alpha} A_\lambda$ is obvious.

If L is dense, then $\alpha = \bigvee_{\lambda < \alpha} \lambda$ (Exercise 36 in Chapter 1). The desired result follows from Proposition 2.13(1) again.

Theorem 2.11. *If L is a complete lattice and $A \in F_L(X)$, then*

(1) $A(x) = \bigvee_{\alpha \in L} (\alpha \wedge A_\alpha(x))$,

(2) $A(x) \geq \bigvee_{\alpha \in L} (\alpha \wedge A_{\dot\alpha}(x))$.

If, furthermore, L is dense, then $A(x) = \bigvee_{\alpha \in L} (\alpha \wedge A_{\dot\alpha}(x))$.

Proof

(1) $\bigvee_{\alpha \in L} (\alpha \wedge A_\alpha(x)) = (\bigvee_{x \in A_\alpha} (\alpha \wedge A_\alpha(x))) \vee (\bigvee_{x \in (A_\alpha)^c} (\alpha \wedge A_\alpha(x)))$
$= \bigvee_{\alpha \leq A(x)} \alpha = A(x)$.

(2) Similar to the proof of (1), we have

$$\bigvee_{\alpha \in L} (\alpha \wedge A_{\dot\alpha}(x)) = \bigvee_{\alpha < A(x)} \alpha \leq A(x).$$

If L is dense, then $\bigvee_{\alpha \in L} (\alpha \wedge A_{\dot\alpha}(x)) = \bigvee_{\alpha < A(x)} \alpha = A(x)$.

Theorem 2.11 can be regarded as the decomposition theorem of an L-fuzzy set. By applying this theorem, it is easy to prove the following:

(1) If L is a complete lattice, then $A = B$ iff $\forall \alpha \in L$, $A_\alpha = B_\alpha$.
(2) If L is a dense complete lattice, then $A = B$ iff $\forall \alpha \in L$, $A_{\dot{\alpha}} = B_{\dot{\alpha}}$.

2.5.4 Mathematical Representation of L-Fuzzy Sets

Suppose that L is a complete lattice. To state the representation theorem of L-fuzzy sets, we firstly introduce a class $\Phi_L(X)$ consisting of those mappings A from L to $P(X)$ satisfying:

(1) $A(0) = X$;
(2) $A(\bigvee_{i \in I} \alpha_i) = \bigcap_{i \in I} A(\alpha_i)$ $(\alpha_i \in L)$.

For $A \in F_L(X)$, define $H : L \to P(X)$ by $\forall \alpha \in L$, $H(\alpha) = A_\alpha$. Then $H \in \Phi_L(X)$. If $H \in \Phi_L(X)$, it can be seen that $\alpha_1 < \alpha_2$ implies that $H(\alpha_1) \supseteq H(\alpha_2)$.

Now we define the operations $\cup$ and $\cap$ on $\Phi_L(X)$.

Let $A_i \in \Phi_L(X)$ $(i \in I)$. Define $\bigcap_{i \in I} A_i$ and $\bigcup_{i \in I} A_i$ by

$$\forall \alpha \in L, \quad (\bigcap_{i \in I} A_i)(\alpha) = \bigcap_{i \in I} A_i(\alpha),$$

$$(\bigcup_{i \in I} A_i)(\alpha) = \bigcap \{A(\alpha) | \forall i \in I, A_i \subseteq A, A \in \Phi_L(X)\},$$

where $A_i \subseteq A$ means that $\forall \alpha \in L$, $A_i(\alpha) \subseteq A(\alpha)$.

Proposition 2.14. $\bigcap_{i \in I} A_i \in \Phi_L(X)$ and $\bigcup_{i \in I} A_i \in \Phi_L(X)$.

Proof. We prove that $\bigcup_{i \in I} A_i \in \Phi_L(X)$, and the proof of $\bigcap_{i \in I} A_i \in \Phi_L(X)$ is left to the reader.

Firstly,

$$(\bigcup_{i \in I} A_i)(0) = \bigcap \{A(0) | \forall i \in I, A_i \subseteq A, A \in \Phi_L(X)\} = X.$$

Next, $\forall \lambda_j \in L$ $(j \in J)$,

$$(\bigcup_{i \in I} A_i)(\bigvee_{j \in J} \lambda_j) = \bigcap \{A(\bigvee_{j \in J} \lambda_j) | \forall i \in I, A_i \subseteq A, A \in \Phi_L(X)\}$$

$$= \bigcap \{\bigcap_{j \in J} A(\lambda_j) | \forall i \in I, A_i \subseteq A, A \in \Phi_L(X)\}$$

$$= \bigcap_{j \in J} \bigcap \{A(\lambda_j) | \forall i \in I, A_i \subseteq A, A \in \Phi_L(X)\}$$

$$= \bigcap_{j \in J} (\bigcup_{i \in I} A_i)(\lambda_j). \qquad \square$$

The above proposition indicates that $\cup$ and $\cap$ are indeed operations on $\Phi_L(X)$. Hence we have the algebraic system $(\Phi_L(X), \cup, \cap)$. It is easily checked that this algebraic system is a complete lattice.

Lemma 2.5. *In a complete lattice* $(L, \vee, \wedge)$,

$$\bigvee_{i\in I} a_i = \bigwedge \{a | \forall i \in I, a_i \leq a\}.$$

Proof. By the definition of supremum, the proof is trivial. $\square$

Lemma 2.6. *If f is a bijection between the complete lattices $(L_1, \vee, \wedge)$ and $(L_2, \vee, \wedge)$ and if f is infimum-preserving, i.e.*

$$f(\bigwedge_{i\in I} a_i) = \bigwedge_{i\in I} f(a_i)(\forall a_i \in L_1),$$

then f is supremum-preserving, i.e.

$$f(\bigvee_{i\in I} a_i) = \bigvee_{i\in I} f(a_i)(\forall a_i \in L_1).$$

Proof. We firstly show that $a \leq b$ iff $f(a) \leq f(b)$.
Indeed,

$$a \leq b \Leftrightarrow a \wedge b = a \Leftrightarrow f(a \wedge b) = f(a) \wedge f(b) = f(a) \Leftrightarrow f(a) \leq f(b).$$

By Lemma 2.5,

$$f(\bigvee_{i\in I} a_i) = f(\bigwedge \{a | \forall i, a_i \leq a\})$$

$$= \bigwedge \{f(a) | a_i \leq a\}$$

$$= \bigwedge \{f(a) | f(a_i) \leq f(a)\}$$

$$= \bigvee_{i\in I} f(a_i)(\forall a_i \in L_1). \quad \square$$

To present the representation theorem, we introduce the following property for a complete lattice $(L, \vee, \wedge)$:

$$(\forall \alpha_i \in L(i \in I)), (\alpha < \bigvee_{i\in I} \alpha_i \Rightarrow \exists i \in I \text{ such that } \alpha \leq \alpha_i),$$

which will be called the condition $(\mathcal{L})$. For instance, a totally ordered set satisfies this condition.

Theorem 2.12. *Let $(L, \vee, \wedge)$ be a complete lattice satisfying the condition $(\mathcal{L})$. If $f : F_L(X) \to \Phi_L(X)$ is defined by*

2.5 L-Fuzzy Sets

$$\forall A \in F_L(X), \forall \alpha \in L, (f(A))(\alpha) = A_\alpha,$$

then f is an isomorphism between $(F_L(X), \cup, \cap)$ and $(\Phi_L(X), \cup, \cap)$.

Proof. We start with the proof that f is surjective. Let $H \in \Phi_L(X)$. For every $x \in X$, define

$$I_x = \{\beta | \beta \in L, x \in H(\beta)\} \text{ and } \beta_0 = \sup I_x.$$

We show that

$$I_x \supseteq \{\beta | 0 \leq \beta \leq \beta_0\} \triangleq [0, \beta_0].$$

If $\beta \in [0, \beta_0[$, then $\beta < \beta_0$. By the condition $(\mathcal{L})$, there exists $\beta_1 \in I_x$ such that $\beta_1 \geq \beta$. Thus $x \in H(\beta_1) \subseteq H(\beta)$, i.e. $x \in H(\beta)$. Consequently, $\beta \in I_x$, which implies $[0, \beta_0[\subseteq I_x$. Since $H \in \Phi_L(X)$,

$$H(\beta_0) = H(\sup I_x) = \bigcap_{\beta \in I_x} H(\beta).$$

When $\beta \in I_x$, $x \in H(\beta)$. Therefore, $x \in H(\beta_0)$, which indicates $\beta_0 \in I_x$. In summary, $[0, \beta_0] \subseteq I_x$. Observing that $[0, \beta_0] \supseteq I_x$, we have $[0, \beta_0] = I_x$.

Now define the L-fuzzy set A by: $\forall x \in X$, $A(x) = \sup I_x$. To prove that f is surjective, it suffices to show that $\forall \alpha \in L$, $H(\alpha) = A_\alpha$. The following equivalencies ensure the desired equality.

$$x \in A_\alpha \Leftrightarrow A(x) \geq \alpha \Leftrightarrow \sup I_x \geq \alpha \Leftrightarrow \beta_0 \geq \alpha \Leftrightarrow \alpha \in I_x \Leftrightarrow x \in H(\alpha).$$

Next we show that f is an injective mapping.

Indeed, if $A, B \in F_L(X)$ satisfy that $f(A) = f(B)$, then

$$\forall \alpha \in L, (f(A))(\alpha) = (f(B))(\alpha),$$

i.e. $A_\alpha = B_\alpha$, and thus $A = B$.

Finally we show that f is intersection-preserving. For $A_i \in F_L(X)$ $(i \in I)$ and $\alpha \in L$,

$$(f(\bigcap_{i \in I} A_i))(\alpha) = (\bigcap_{i \in I} A_i)_\alpha = \bigcap_{i \in I}(A_i)_\alpha$$
$$= \bigcap_{i \in I}(f(A_i))(\alpha) = (\bigcap_{i \in I} f(A_i))(\alpha),$$

or equivalently,

$$f(\bigcap_{i \in I} A_i) = \bigcap_{i \in I} f(A_i).$$

That f is union-preserving follows from Lemma 2.6. □

Remark 2.9. *Besides Theorem 2.12, more isomorphism theorems can be found in [74] which reveal the relationships between the set of L-fuzzy sets and other different classes of sets.*

Finally, we turn to a special case $\Phi_{[0,1]}(X)$ of $\Phi_L(X)$. For convenience, we simply write $\Phi(X)$ instead of $\Phi_{[0,1]}(X)$. Then a direct application to $\Phi(X)$ of Theorem 2.12 leads to the following conclusion.

Corollary 2.4. *The algebraic systems* $(F(X), \cup, \cap)$ *and* $(\Phi(X), \cup, \cap)$ *are isomorphic.*

In addition, we can define a unary operation c in $\Phi(X)$.

$$\forall F \in \Phi(X), \forall \alpha \in [0,1], F^c(\alpha) = \bigcap_{\lambda < \alpha} (F(1-\lambda))^c.$$

Since $\forall \alpha_i \in [0,1]$ $(i \in I)$,

$$F^c(\bigvee_{i \in I} \alpha_i) = \bigcap_{\lambda < \bigvee_{i \in I} \alpha_i} (F(1-\lambda))^c$$
$$= \bigcap_{i \in I} \bigcap_{\lambda < \alpha_i} (F(1-\lambda))^c$$
$$= \bigcap_{i \in I} F^c(\alpha_i),$$

we have $F^c \in \Phi(X)$. Furthermore, if the mapping f from $F(X)$ to $\Phi(X)$ is defined as in Theorem 2.12, then $\forall A \in F(X), \forall \alpha \in [0,1]$,

$$(f(A))^c(\alpha) = \bigcap_{\lambda < \alpha} ((f(A))(1-\lambda))^c$$
$$= \bigcap_{\lambda < \alpha} (A_{1-\lambda})^c = (\bigcup_{\lambda < \alpha} A_{1-\lambda})^c$$
$$= (\bigcup_{\lambda' > 1-\alpha} A_{\lambda'})^c = (A_{1-\alpha})^c \quad \text{(by Proposition 2.8(4))}$$
$$= (A^c)_\alpha = (f(A^c))(\alpha).$$

Thus, f preserves complement. In summary, we come to the following conclusion.

Theorem 2.13. *If* $f : F(X) \to \Phi(X)$ *is defined by* $\forall A \in F(X), \forall \alpha \in [0,1]$, $(f(A))(\alpha) = A_\alpha$, *then* f *is an isomorphism between* $(F(X), \cup, \cap, c)$ *and* $(\Phi(X), \cup, \cap, c)$.

2.6 Fuzzy Pattern Recognition

By pattern, we mean prototypes which serve as standard objects. As its name suggests, pattern recognition is to match a given object with one or more of these standard objects. Broadly speaking, we recognize objects every day.

2.6 Fuzzy Pattern Recognition

While we look for a toothbrush, a cup, a bag, or while we see a sight, read a letter or process an image, we are carrying out pattern recognition. However, our use of the term refers especially to the recognition by means of mathematical approaches under the condition that both pattern and objects to be recognized could be mathematically characterized. In pattern recognition, the involved pattern or objects are frequently fuzzy, i.e. they have vague boundaries. For instance, one often describes a person by using words such as middle-aged, in a south accent, very short and fat etc. Based on these fuzzy descriptions, we can find the right person. This process is just an example of fuzzy pattern recognition. In this section, we shall introduce an application of fuzzy set theory to two types of fuzzy pattern recognition models.

2.6.1 Type I Fuzzy Pattern Recognition

Let A_1, A_2, ..., A_n be n fuzzy sets on X representing n standard classes. Given $x_0 \in X$, we need to know which class x_0 should belong to.

The rule of maximum membership degree
If $A_k(x_0) = \max\{A_1(x_0), A_2(x_0), \ldots, A_n(x_0)\}$ ($1 \leq k \leq n$), then x_0 should be classified into the k-th class A_k.

Example 2.17. $O =$old, $Y =$young, $M =$middle-aged, $X = [0, 100]$.

$$O(x) = \begin{cases} [1 + (\frac{x-50}{5})^{-2}]^{-1} & \text{if } 50 < x \leq 100 \\ 0 & \text{otherwise} \end{cases},$$

$$Y(x) = \begin{cases} [1 + (\frac{x-25}{5})^2]^{-1} & \text{if } 25 < x \leq 100 \\ 1 & \text{otherwise} \end{cases},$$

$M = O^c \cap Y^c$.

If $x_0 = 35$, then $O(x_0) = 0$, $Y(x_0) = 0.2$, $M(x_0) = 0.8$. According to the rule of maximum membership degree, x_0 should be classified into M, i.e. a person of 35 years old should be classified as middle aged.

Example 2.18. Let $X = \{(A, B, C) | A \geq B \geq C, A+B+C = 180\}$ represent the set of triangles.

Approximately right triangles R:

$$R(x) = R(A, B, C) = 1 - \frac{1}{90}|A - 90|.$$

Approximately isosceles triangles I:

$$I(x) = I(A, B, C) = 1 - \frac{1}{60} \min((A - B), (B - C)).$$

Approximately equilateral triangles E:

$$E(x) = E(A,B,C) = 1 - \frac{1}{180}(A-C).$$

Other triangles T: $T = R^c \cap I^c \cap E^c$.

Given a triangle $x_0 = (A,B,C) = (80, 55, 45)$, we have $R(x_0) = 0.87$, $I(x_0) = 0.83$, $E(x_0) = 0.81$, $T(x_0) = 1 - \max(R(x_0), I(x_0), E(x_0)) = 0.13$. According to the rule of maximum membership degree, a triangle with three angles 80,55,45 can be regarded as an approximately right triangle.

In some cases, an object may have very close membership degrees to some fuzzy standard classes that the following rule may be more appropriate.

The rule of threshold

Given a threshold $\alpha \in [0,1]$, if $A_{i_1}(x_0) \geq \alpha$, $A_{i_2}(x_0) \geq \alpha$, ..., $A_{i_k}(x_0) \geq \alpha$, then x_0 should be classified into the k classes $A_{i_1}, A_{i_2}, \ldots, A_{i_k}$ simultaneously.

For instance, in Example 2.18, let $x_0 = (85, 50, 45)$. Then $R(x_0) = 0.95$, $I(x_0) = 0.91$, $E(x_0) = 0.87$, $T(x_0) = 0.05$. If we take $\alpha = 0.9$, the triangle x_0 should be in the class of approximately isosceles right triangles.

2.6.2 Type II Fuzzy Pattern Recognition

In the previous subsection, the recognition issue is of the following characteristic: the standard classes are fuzzy and the object (an element in X) to be recognized is crisp. In this subsection, we pay attention to the issue with fuzzy standard classes and fuzzy object. More precisely, let $A_1, A_2, \ldots, A_n$ be n fuzzy sets on X for n standard classes. Given $A \in F(X)$, we need to know which class A should be identified with. To solve this problem, we need to measure how close two fuzzy sets are.

Definition 2.17. *If $N : F(X) \times F(X) \to [0,1]$ satisfies that*

(1) $N(\emptyset, X) = 0$ and $N(A,A) = 1$ whenever $A \in F(X)$,
(2) $N(A,B) = N(B,A)$ whenever $A,B \in F(X)$,
(3) $N(A,C) \leq \min(N(A,B), N(B,C))$ whenever $A \subseteq B \subseteq C$,

then N is called a nearness measure.

Given a nearness measure N, $N(A,B)$ measures the nearness of the two fuzzy sets A and B. Now we list some specific nearness measures for practical use.

The first type of nearness measures is defined by means of some distance or metric.

Let $X = \{x_1, x_2, \ldots, x_n\}$ and $A, B \in F(X)$. If the Hamming distance is chosen, we define

2.6 Fuzzy Pattern Recognition

$$N_H(A,B) = 1 - \frac{1}{n}\sum_{i=1}^{n}|A(x_i) - B(x_i)|.$$

It is easily checked that N_H meets the requirements in Definition 2.17, and thus constitutes a nearness measure.

For a continuous universe $X = [a,b]$ and $A, B \in F(X)$, the nearness measure with the Hamming distance is defined by

$$N_H(A,B) = 1 - \frac{1}{b-a}\int_a^b |A(x) - B(x)|dx,$$

provided the integral exists.

Similarly, the nearness measure with the Euclidean distance is defined by

$$N_E(A,B) = 1 - \frac{1}{\sqrt{n}}\left(\sum_{i=1}^{n}(A(x_i) - B(x_i))^2\right)^{\frac{1}{2}}$$

for a discrete universe $X = \{x_1, x_2, \ldots, x_n\}$ and $A, B \in F(X)$;

$$N_E(A,B) = 1 - \frac{1}{\sqrt{b-a}}\left(\int_a^b (A(x) - B(x))^2 dx\right)^{\frac{1}{2}}$$

for a continuous universe $X = [a,b]$ and $A, B \in F(X)$ provided the integral exists.

Remark 2.10. *More generally, Minkowski distance can also be used for constructing a nearness measure. However, the above mentioned distance-related nearness measures are enough for practical use.*

The second type of nearness measures includes the following.

(1) Given a discrete universe $X = \{x_1, x_2, \ldots, x_n\}$, $A, B \in F(X)$,

$$N_{M_1}(A,B) = \frac{\sum_{i=1}^{n} A(x_i) \wedge B(x_i)}{\sum_{i=1}^{n} A(x_i) \vee B(x_i)}.$$

For a continuous universe $X = [a,b]$,

$$N_{M_1}(A,B) = \frac{\int_a^b (A(x) \wedge B(x))dx}{\int_a^b (A(x) \vee B(x))dx},$$

provided the integral exists.

(2) Given a discrete universe $X = \{x_1, x_2, \ldots, x_n\}$, $A, B \in F(X)$,

$$N_{M_2}(A,B) = \frac{2\sum_{i=1}^{n} A(x_i) \wedge B(x_i)}{\sum_{i=1}^{n} A(x_i) + B(x_i)}.$$

For a continuous universe $X = [a, b]$,

$$N_{M_2}(A, B) = \frac{2 \int_a^b (A(x) \wedge B(x))dx}{\int_a^b (A(x) + B(x))dx},$$

provided the integral exists.

The proof that both N_{M_1} and N_{M_2} are nearness measures is left to the reader.

The last one is the so-called lattice nearness measure which is defined by means of the inner and outer product of two fuzzy sets.

Definition 2.18. *For $A, B \in F(X)$,*

$$A \odot B = \sup_{x \in X}(A(x) \wedge B(x)) = hgt(A \cap B)$$

is called the inner product of A and B.

$$A \otimes B = \inf_{x \in X}(A(x) \vee B(x))$$

is called the outer product of A and B.

With the notion of inner and outer product, we define

$$\forall A, B \in F(X), N_L(A, B) = (A \odot B) \wedge (1 - A \otimes B).$$

The N_L has the following properties:

(1) $N_L(\emptyset, X) = 0$;
(2) $N_L(A, A) = \bar{a} \wedge (1 - \underline{a})$, where $\bar{a} = hgt(A)$ and $\underline{a} = plt(A)$;
(3) $N_L(A, B) = N_L(B, A)$;
(4) $A \subseteq B \subseteq C \Rightarrow N_L(A, C) \leq \min(N_L(A, B), N_L(B, C))$.

The proofs of these properties are straightforward and omitted here.

These properties indicate that N_L is not a nearness measure in strict sense since $N_L(A, A) = 1$ does not necessarily hold. However, it is an important index in evaluating the nearness of two fuzzy sets and extensively used in the literature. In addition, it follows from (3) that $N_L(A, A) = 1$ iff $\bar{a} = 1$ and $\underline{a} = 0$, which are true for fuzzy sets in practice. Therefore, we keep it in the family of nearness measures and call it the lattice nearness measure.

Example 2.19. *Let $X = \mathbb{R}$ and $A_i(x) = e^{-\left(\frac{x - a_i}{\sigma_i}\right)^2}$ $(\sigma_i > 0; i = 1, 2)$. Find $N_L(A_1, A_2)$.*

Solution. *Clearly, $A_1 \otimes A_2 = 0$.*

$$A_1 \odot A_2 = \sup_{x \in \mathbb{R}} A_1(x) \wedge A_2(x) = A_1(x^*) = A_2(x^*),$$

2.6 Fuzzy Pattern Recognition

where x^* is between a_1 and a_2. It follows from $A_1(x^*) = A_2(x^*)$ that

$$x_1^* = \frac{\sigma_1 a_2 + \sigma_2 a_1}{\sigma_1 + \sigma_2} \qquad x_2^* = \frac{\sigma_2 a_1 - \sigma_1 a_2}{\sigma_2 - \sigma_1}.$$

x_2^* is removed since it is not between a_1 and a_2. As a consequence,

$$x^* = x_1^* = \frac{\sigma_1 a_2 + \sigma_2 a_1}{\sigma_1 + \sigma_2}.$$

Therefore, $A_1 \odot A_2 = e^{-\left(\frac{a_2 - a_1}{\sigma_1 + \sigma_2}\right)^2}$, whence

$$N_L(A_1, A_2) = (A_1 \odot A_2) \wedge (1 - A_1 \otimes A_2) = A_1 \odot A_2 = e^{-\left(\frac{a_2 - a_1}{\sigma_1 + \sigma_2}\right)^2}.$$

Particularly, $a_1 = a_2 \Rightarrow N_L(A_1, A_2) = 1$.

We see from Example 2.19 that the application of N_L places much emphasis upon the relative position of peak points of two fuzzy sets.

Now let us return to the type II fuzzy pattern recognition model. Assume that $A_1, A_2, \ldots, A_n \in F(X)$ are n patterns and $A \in F(X)$ is the object to be recognized. Then we can choose a nearness measure N and compute $N(A, A_i)$ $(i = 1, 2, \ldots, n)$. If $N(A, A_k) = \max(N(A, A_1), N(A, A_2), \ldots, N(A, A_n))$ $(1 \leq k \leq n)$, then A is identified with the pattern A_k.

Example 2.20. Let $X =]0, +\infty[$ be the weight range of 100 grains of wheat with distribution $A(x) = e^{-\left(\frac{x-a}{\sigma}\right)^2}$. The types of wheat and the corresponding parameters for each type of wheat are listed in the following.

type	A_1	A_2	A_3	A_4	A_5
a	3.7	2.9	5.6	3.9	3.7
σ	0.3	0.3	0.3	0.3	0.2

Now comes the wheat with distribution $A(x) = e^{-\left(\frac{x-3.43}{0.28}\right)^2}$. The question is: what is the type of the wheat?

Solution. Choose N_L as the nearness measure of fuzzy sets. Then, by the formula in Example 2.19, we have

$$N_L(A, A_1) = e^{-\left(\frac{3.7 - 3.43}{0.28 + 0.3}\right)^2} \approx 0.78.$$

By similar computations, we have

$$N_L(A, A_2) \approx 0.44, \ N_L(A, A_3) \approx 0, \ N_L(A, A_4) \approx 0.52, \ N_L(A, A_5) \approx 0.68.$$

Since $N_L(A, A_1) > N(A, A_i)$ $(i = 2, 3, 4, 5)$, hence A should be recognized as A_1.

2.7 Exercises

1. Assume two fuzzy sets A_1 and A_2 on $X = \{x_1, x_2, x_3, x_4\}$ are defined by

$$A_1 = 0.1/x_1 + 0.9/x_2 + 0.6/x_3,$$
$$A_2 = 0.9/x_1 + 0.7/x_2 + 0.6/x_3 + 0.8/x_4.$$

   Find A_1^c, $A_1 \cup A_2$ and $A_1 \cap A_2$.

2. If $A, B, C \in F(X)$, show that

   (1) $(A \cap ((B \cap C) \cup (A^c \cap C^c))) \cup C^c = (A \cap B \cap C) \cup C^c$,
   (2) $(A \cap B) \cup (B \cap C) \cup (A \cap C) = (A \cup B) \cap (B \cup C) \cap (A \cup C)$.

3. The difference $A - B$ and symmetric difference $A \triangle B$ of two fuzzy sets A and B are respectively defined by $A - B = A \cap B^c$ and $A \triangle B = (A - B) \cup (B - A)$.

   (i) Use $A(x)$ and $B(x)$ to express $(A - B)(x)$ and $(A \triangle B)(x)$.
   (ii) Assume A and B are two fuzzy sets on $X = \{a, b, c, d, e, f, g\}$ defined by

$$A = 0.5/b + 0.4/c + 1/d + 0.7/f,$$
$$B = 0.3/a + 0.9/b + 0.4/c + 1/d + 0.6/e + 1/g.$$

   Find $A - B$ and $A \triangle B$.
   (iii) Show that $(A \triangle B) \triangle C = A \triangle (B \triangle C)$.

4. Show that the relation $\leq$ in the soft algebra $(F(X), \cup, \cap, c)$ is $\subseteq$.

5. Prove the following identities: $\forall A, A_i, B_i \in F(X)$ ($i \in I$),

   (1) $(\bigcup_{i \in I} A_i) \cup (\bigcup_{i \in I} B_i) = \bigcup_{i \in I} (A_i \cup B_i)$ and
   $(\bigcap_{i \in I} A_i) \cap (\bigcap_{i \in I} B_i) = \bigcap_{i \in I} (A_i \cap B_i)$;
   (2) $A \cup (\bigcap_{i \in I} A_i) = \bigcap_{i \in I} (A \cup A_i)$ and $A \cap (\bigcup_{i \in I} A_i) = \bigcup_{i \in I} (A \cap A_i)$;
   (3) $(\bigcup_{i \in I} A_i)^c = \bigcap_{i \in I} (A_i)^c$ and $(\bigcap_{i \in I} A_i)^c = \bigcup_{i \in I} (A_i)^c$.

6. Let $X = \{x_1, x_2, x_3, x_4, x_5\}$ and the fuzzy set A on X be defined by

$$A = 1/x_1 + 0.8/x_2 + 0.3/x_3 + 1/x_4.$$

   Find $\ker(A)$, $\text{supp}(A)$, $\text{hgt}(A)$ and $\text{plinth}(A)$. Is A normal? If $X = \{x_1, x_2, x_3, x_4\}$, find $\text{plinth}(A)$.

7. Prove the following identities: $\forall A, B \in F(X)$,

   (1) $\text{hgt}(A \cup B) = \max\{\text{hgt}(A), \text{hgt}(B)\}$.
   (2) $\text{hgt}(A \cap B) \leq \min\{\text{hgt}(A), \text{hgt}(B)\}$. Give an example to illustrate that the inequality can not be changed into the equality.

(3) hgt$(A \cup A^c) \geq 0.5$ (the weakened law of excluded middle) and hgt$(A \cap A^c) \leq 0.5$ (the weakened law of contradiction).

8. According to your understanding, construct the membership functions of "approximately between 1 and 2" and "evening".
9. Define N_λ ($\lambda > -1$) by $\forall x \in [0,1]$, $N_\lambda(x) = \frac{1-x}{1+\lambda x}$. Show that N_λ ($\lambda > -1$) is a strong negation.
10. If n is a (strict, strong) negation, show that n_d defined by: $\forall x \in [0,1]$, $n_d(x) = 1 - n(1-x)$ is a (strict, strong) negation as well.
11. Show that a decreasing function $n: [0,1] \to [0,1]$ satisfying involution is a strong negation.
12. Let N be a strong negation satisfying: $\forall x \in [0,1]$, $N(x) = 1 - N(1-x)$. Show that N is the standard negation.
13. Prove Lemma 2.2 and the representation theorem of a strong negation.
14. Show that both φ and ψ are two generators of the same strong negation iff there exists an automorphism η on $[-\frac{1}{2}, \frac{1}{2}]$ such that $\forall x \in [-\frac{1}{2}, \frac{1}{2}]$,
$$\psi(x) = 1/2 + \eta(\varphi(x) - 1/2) \text{ and } \eta(-x) = -\eta(x).$$

15. Define
$$T(x,y) = \frac{xy}{\lambda + (1-\lambda)(x+y-xy)}$$
and
$$S(x,y) = \frac{x+y-xy-(1-\lambda)xy}{1-(1-\lambda)xy}.$$
Show that T and S are a t-norm and a t-conorm respectively for $\lambda > 0$.
16. Verify that T^φ and S^φ are a t-norm and a t-conorm respectively.
17. Show that the only t-norm T satisfying $T(x,x) = 0$ for all $x \in [0,1[$ is T_0.
18. If $f: [0,1] \to [0,+\infty[$ is a strictly decreasing continuous function with $f(1) = 0$ and $T(x,y) = f^{-1}(f(0) \wedge (f(x)+f(y)))$, then T is a continuous t-norm satisfying: $\forall x \in]0,1[$, $T(x,x) < x$ (Archimedean t-norm).
19. Assume two fuzzy sets A_1 and A_2 on $X = \{x_1, x_2, x_3, x_4\}$ are defined by
$$A_1 = 0.1/x_1 + 0.9/x_2 + 0.6/x_3,$$
$$A_2 = 0.3/x_1 + 0.7/x_2 + 0.2/x_3 + 0.8/x_4.$$
Find $A_1 \cup_{S_L} A_2$, $A_1 \cap_{T_L} A_2$ and $A_1 \cup_{S_\pi} A_2^c$, $A_1^c \cap_{T_\pi} A_2$.
20. Show that $A \cup_{S_L} A^c = X$ and $A \cap_{T_L} A^c = \emptyset$ for all $A \in F(X)$, i.e. the law of excluded middle and the law of contradiction still hold provided that the Lukasiewicz t-norm and t-conorm are chosen to represent union and intersection respectively.
21. Assume that (T, S, n) be a De Morgan triple. If n is a strong negation with a generator φ, show that:

(1) If $T = T_L^\varphi$, then $S = S_L^\varphi$;
(2) If $T = T_\pi^\varphi$, then $S = S_\pi^\varphi$.

22. Let T be a t-norm and a, a_i ($i \in I$) be real numbers in $[0,1]$. Show that $T(\sup_{i\in I} a_i, a) = \sup_{i\in I} T(a_i, a)$ iff T is left continuous.

23. Let T be a t-norm such that $\forall x, y \in [0,1]$, $T(x,y) + T(x, 1-y) = x$. Show that $T = T_\pi$.

24. A mapping U from $[0,1] \times [0,1]$ to $[0,1]$ is called a uninorm if it satisfies symmetry, monotonicity, associativity and the existence of identity, i.e. there exists an $e \in [0,1]$ such that $T(e, x) = x$ whenever $x \in [0,1]$. Verify that every t-norm or t-conorm is a uninorm. Let c be a number in $[0,1]$. Define U_1 and U_2 respectively by $\forall x, y \in [0,1]$

$$U_1(x,y) = \begin{cases} x \vee y & x \wedge y > c \\ x \wedge y & \text{otherwise} \end{cases}$$

and

$$U_2(x,y) = \begin{cases} x \wedge y & x \vee y < c \\ x \vee y & \text{otherwise} \end{cases}.$$

Show that both U_1 and U_2 are uninorms with identity c.

25. If I is a fuzzy implication, show that $n(x) = I(x,0)$ is a fuzzy negation.

26. Let φ be an automorphism and T a t-norm. Define T^φ by: $\forall x, y \in [0,1]$,

$$T^\varphi(x,y) = \varphi^{-1}(T(\varphi(x), \varphi(y))).$$

Find $I_{T_L^\varphi}$ and $I_{T_\pi^\varphi}$.

27. Show that every R-implication I_T satisfies that $\forall x, y \in [0,1]$, $x \leq y$ implies $I_T(x,y) = 1$.

28. If a t-norm T is positive, i.e. $\forall x, y \in]0,1]$, $T(x,y) > 0$, show that $n(x) = I_T(x,0)$ is the intuitionistic negation.

29. If T is a left continuous t-norm, show that

$$\forall x, y \in [0,1], T(x, I_T(x,y)) = \min\{x, y\}.$$

30. Show that the Lukasiewice equivalence E can be written as $E(x,y) = \min(I_{T_L}(x,y), I_{T_L}(y,x))$.

31. Let $X = \{x_1, x_2, x_3, x_4, x_5, x_6, x_7\}$ and a fuzzy set A on X be defined by

$$A = 0.2/x_1 + 0.1/x_2 + 0.8/x_3 + 0.6/x_4 + 0.4/x_5 + 0.6/x_6 + 0.9/x_7.$$

Find $A_{0.4}, A_{0.6}, A_{\dot{0.4}}, A_{\dot{0.6}}$.

32. Given a fuzzy set A on $\mathbb{R}$ with the membership function $A(x) = e^{-x^2}$, find $A_{1/e}, A_{\dot{1/e}}, A_1$ and $A_{\dot{0}}$.

33. Let $A, B \in F(X)$. Show that $A \subseteq B$ iff $\forall \alpha \in [0,1]$, $A_\alpha \subseteq B_\alpha$.

2.7 Exercises

34. If the α-cuts of a fuzzy set A on $X = \{a, b, c, d, e\}$ are

$$A_\alpha = \begin{cases} \{d\} & 0.7 < \alpha \leq 1 \\ \{c, d\} & 0.5 < \alpha \leq 0.7 \\ \{c, d, e\} & 0.3 < \alpha \leq 0.5 \\ \{b, c, d, e\} & 0.1 < \alpha \leq 0.3 \\ X & 0 \leq \alpha \leq 0.1 \end{cases},$$

find the fuzzy set A.

35. Prove the following identities:

(1) $A_{\underset{\sim}{\alpha}} = \bigcup_{\lambda > \alpha} A_\lambda, \quad A_\alpha = \bigcap_{\lambda < \alpha} A_\lambda;$

(2) $(\bigcup_{i \in I} A_i)_\alpha = \bigcap_{\lambda < \alpha} \bigcup_{i \in I} (A_i)_\lambda, \quad (\bigcap_{i \in I} A_i)_{\underset{\sim}{\alpha}} = \bigcup_{\lambda > \alpha} \bigcap_{i \in I} (A_i)_\lambda.$

36. Let $X = [0, 5]$. Find the fuzzy set A whose α-cuts are

$$A_\alpha = \begin{cases} [0, 5] & \alpha = 0 \\ (3\alpha, 5] & 0 < \alpha \leq 2/3 \\ (3, 5] & 2/3 < \alpha \leq 1 \end{cases}$$

37. Let $r_1, r_2, r_3, \ldots$ be all the rational numbers in $[0, 1]$ and $A, B \in F(X)$. Prove that

(i) $A = \bigcup_{n=1}^{\infty} r_n A_{r_n};$

(ii) $A = B$ iff $A_{r_i} = B_{r_i}$ for $i = 1, 2, 3, \ldots$.

38. Let H be a nest of sets on $X = [-1, 1]$ defined by $H(\alpha) = [\alpha^2 - 1, 1 - \alpha^2]$ ($\forall \alpha \in [0, 1]$). Determine the membership function of the fuzzy set $A = \bigcup_{\alpha \in [0,1]} \alpha H(\alpha)$ on X.

39. Use Representation Theorem to show that $\forall A, B \in F(X)$, $(A \cup B)^c = A^c \cap B^c$.

40. Give an example to illustrate that $(U(X), \cup, \cap, c)$ is not a Boolean algebra.

41. Let N be a pseudo-complement on a poset $(P, \leq)$. If $\alpha < \beta$, show that $N(\beta) < N(\alpha)$.

42. Let L be a dense complete lattice and $A \in F_L(X)$. If there exists a mapping $H : L \to P(X)$ such that $A_{\underset{\sim}{\alpha}} \subseteq H(\alpha) \subseteq A_\alpha$ ($\forall \alpha \in L$), prove that $A(x) = \sup\{\alpha | x \in H(\alpha)\}$.

43. Let L be a complete lattice satisfying the property: $\forall \alpha_i \in L (i \in I), \exists i_0 \in I$ such that $\alpha_{i_0} = \bigvee_{i \in I} \alpha_i$. If there exists a mapping $H : L \to P(X)$ such that (i) $\alpha < \beta$ implies that $H(\beta) \subseteq H(\alpha)$; (ii) $A(x) = \sup\{\alpha | x \in H(\alpha)\}$, prove that $A_{\underset{\sim}{\alpha}} \subseteq H(\alpha) \subseteq A_\alpha$ ($\forall \alpha \in L$).

44. Let L be a complete lattice and $F \in \Phi_L(X)$. If $A(x) = \bigvee_{\alpha \in L}(\alpha \wedge F(\alpha)(x))$, show that $\forall \alpha \in L, A_\alpha = F(\alpha)$.

45. A mapping $f : [0,1] \to P(R)$ is defined by
$$f(\alpha) = \begin{cases} \emptyset & \alpha = 1 \\ [\alpha, 2-\alpha] & \alpha \neq 0, 1 \\ [-2, 2] & \alpha = 0. \end{cases}$$
Prove that $f(\alpha) \subseteq f(\beta)$ $(\alpha > \beta)$ and $f(\bigvee_{\alpha<1} \alpha) \neq \bigcap_{\alpha<1} f(\alpha)$.

46. Given $X = \{x_1, x_2, x_3\}$, $A = 0.7/x_1 + 0.6/x_2 + 0.4/x_3$, $B = 0.5/x_1 + 0.6/x_3$, $C = 0.6/x_1 + 0.8/x_2 + 0.3/x_3$, use the lattice nearness measure N_L to determine which two fuzzy sets are the nearest. If the nearness measure N_H is employed, is the conclusion the same?

47. Show that the inner product and outer product have the following properties:

    (1) $A \odot B = B \odot A$ and $A \otimes B = B \otimes A$;
    (2) $\sup\{A \odot B | B \in F(X)\} = A \odot A = hgt(A)$ and $\inf\{A \otimes B | B \in F(X)\} = A \otimes A$;
    (3) $(A \cup B) \odot C = (A \odot C) \vee (B \odot C)$ and $(A \cap B) \otimes C = (A \otimes C) \wedge (B \otimes C)$.

48. Let $X = [0, 2]$ and $A, B \in F(X)$ be defined by
$$A(x) = \begin{cases} x & 0 \leq x < 1 \\ 2 - x & 1 \leq x \leq 2 \end{cases}$$
$$B(x) = \sqrt{1 - (1-x)^2}.$$
Compute $N_H(A, B)$.

49. In weather forecast, the weather pre-evaluation is defined as $N_L(A, B)$, where $A(x) = e^{-(\frac{x-a_1}{\sigma})^2}$, $B(x) = e^{-(\frac{x-a_2}{\sigma})^2}$, a_1 is the actual amount of rainfall (in: mm), a_2 is the prediction value, and σ is the variance. Now there are two places where the following data are obtained. In one place, the actual value and prediction values are 220mm and 225mm respectively. The corresponding values are 40mm and 50mm respectively in the other place. If the variance $\sigma = 1$, compute the pre-evaluations in the two places.

Chapter 3
Fuzzy Relations

As known to us, a relation is a subset of the Cartesian product of two sets. A relation is naturally fuzzified while a subset is fuzzified. In fact, whether two objects have a relation is not always easy to determine. For example, the relation "greater than" on the set of real numbers is a crisp one because we can determine the order relation of any two real numbers without vagueness. However, the relation "much greater than" is a fuzzy one because it is impossible for us to figure out the exact minimum difference of two numbers satisfying this relation. In real world problems, there exist a lot of such relations, e.g. " being friend of" and "being confident in" between some people. These relations will be termed as fuzzy relations which are our concern in this chapter.

3.1 Basic Concepts of Fuzzy Relations

Definition 3.1. *Let X and Y be two non-empty sets. A mapping $R : X \times Y \to [0,1]$ is called a fuzzy (binary) relation from X to Y. For $(x,y) \in X \times Y$, $R(x,y) \in [0,1]$ is referred to as the degree of relationship between x and y. Particularly, a fuzzy relation from X to X is called a fuzzy (binary) relation on X.*

By definition, a fuzzy relation R is a fuzzy set on $X \times Y$, i.e. $R \in F(X \times Y)$. We know that the relation $>$ (greater than) on the set of real numbers is a crisp relation with the characteristic function defined by

$$> (x,y) = \begin{cases} 1 & x > y; \\ 0 & \text{otherwise.} \end{cases}$$

The relation ">>=much greater than" is a fuzzy relation on the set of real numbers, which may be expressed by

$$>> (x,y) = \begin{cases} (1 + \frac{100}{(x-y)^2})^{-1} & x > y; \\ 0 & \text{otherwise.} \end{cases}$$

For instance, the ordered pairs $(x+1, x)$ have a low degree $\frac{1}{101}$ with respect to ">>", the ordered pairs $(x+10, x)$ have an intermediate degree 0.5 with respect to ">>", and the ordered pairs $(x+100, x)$ have a high degree $\frac{100}{101}$ with respect to ">>".

Let R be a fuzzy relation from X to Y. The R-afterset xR of x $(x \in X)$ is a fuzzy set on Y defined by

$$\forall y \in Y, (xR)(y) = R(x, y).$$

The R-foreset Ry of y $(y \in Y)$ is a fuzzy set on X defined by

$$\forall x \in X, (Ry)(x) = R(x, y).$$

Since fuzzy relations are fuzzy sets, they have the same set-theoretic operations as fuzzy sets. Let R and S be fuzzy relations from X to Y.

R is contained in S, denoted $R \subseteq S$, iff $\forall (x, y) \in X \times Y$, $R(x, y) \leq S(x, y)$; R is equal to S, denoted $R = S$, iff $\forall (x, y) \in X \times Y$, $R(x, y) = S(x, y)$.

Clearly, $R = S$ iff $R \subseteq S$ and $S \subseteq R$.

The union $R \cup S \in F(X \times Y)$ of R and S is defined by

$$\forall (x, y) \in X \times Y, (R \cup S)(x, y) = R(x, y) \vee S(x, y);$$

The intersection $R \cap S \in F(X \times Y)$ of R and S is defined by

$$\forall (x, y) \in X \times Y, (R \cap S)(x, y) = R(x, y) \wedge S(x, y);$$

The complement $R^c \in F(X \times Y)$ of R is defined by

$$\forall (x, y) \in X \times Y, (R^c)(x, y) = 1 - R(x, y);$$

The inverse $R^{-1} \in F(Y \times X)$ of R is defined by

$$\forall (x, y) \in X \times Y, R^{-1}(y, x) = R(x, y).$$

In addition, if $R_i \in F(X \times Y)$ $(i \in I)$, then $\bigcup_{i \in I} R_i$ is defined by

$$\forall (x, y) \in X \times Y, (\bigcup_{i \in I} R_i)(x, y) = \bigvee_{i \in I} R_i(x, y),$$

and $\bigcap_{i \in I} R_i$ is defined by

$$\forall (x, y) \in X \times Y, (\bigcap_{i \in I} R_i)(x, y) = \bigwedge_{i \in I} R_i(x, y).$$

It follows from Theorem 2.2 in Chapter 2 that $(F(X \times Y), \cup, \cap, c)$ is a superior soft algebra. We shall use the properties of a superior soft algebra for fuzzy

3.1 Basic Concepts of Fuzzy Relations

relations without any further interpretations. Moreover, it is easily checked that

(1) $(R \cup S)^{-1} = R^{-1} \cup S^{-1}$;
(2) $(R \cap S)^{-1} = R^{-1} \cap S^{-1}$;
(3) $(R^c)^{-1} = (R^{-1})^c$.

A fuzzy relation also has the concept of (strong) α-cut. The crisp relation

$$R_\alpha = \{(x,y)|R(x,y) \geq \alpha\}(\alpha \in [0,1])$$

will be called the α-cut relation of R, and

$$R_{\dot\alpha} = \{(x,y)|R(x,y) > \alpha\}(\alpha \in [0,1])$$

will be called the strong α-cut relation of R.

Clearly, both an α-cut relation and a strong α-cut relation are crisp relations from X to Y. Naturally, (strong) α-cut relations have all the properties valid for (strong) α-cuts of a fuzzy set, e.g. $(R \cup S)_\alpha = R_\alpha \cup S_\alpha$, $(R^c)_\alpha = (R_{1-\dot\alpha})^c$, $R(x,y) = \bigvee_{\alpha \in [0.1]} (\alpha \wedge R_\alpha(x,y))$ etc..

In the rest of this section, we concentrate on fuzzy relations between finite universes. More specifically, let R be a fuzzy relation from X to Y, where $X = \{x_1, x_2, \cdots, x_n\}$ and $Y = \{y_1, y_2, \cdots, y_m\}$. In this case, by letting $r_{ij} = R(x_i, y_j)$ $(i = 1, 2, \cdots, n; j = 1, 2, \cdots, m)$, the fuzzy relation R may be represented in the form of a matrix

$$\begin{pmatrix} r_{11} & r_{12} & \cdots & r_{1m} \\ r_{21} & r_{22} & \cdots & r_{2m} \\ \cdots & \cdots & \cdots & \cdots \\ r_{n1} & r_{n2} & \cdots & r_{nm} \end{pmatrix}$$

Henceforth, we shall not distinguish between a fuzzy relation R and the corresponding matrix. In other words, we simply write $R = (r_{ij})_{n \times m}$.

Example 3.1. *Given the universe of height* $X = \{140, 150, 160, 170, 180\}$ *(in: cm) and the universe of weight* $Y = \{40, 50, 60, 70, 80\}$ *(in: kg), the relation between the height and weight of a person may be regarded as a fuzzy relation* R *which is expressed as:*

$$R = \begin{pmatrix} 1 & 0.8 & 0.2 & 0.1 & 0 \\ 0.8 & 1 & 0.8 & 0.2 & 0.1 \\ 0.2 & 0.8 & 1 & 0.8 & 0.2 \\ 0.1 & 0.2 & 0.8 & 1 & 0.8 \\ 0 & 0.1 & 0.2 & 0.8 & 1 \end{pmatrix}.$$

Example 3.2. *Suppose* X *represents a set of patients* $X = \{p_1, p_2, p_3, p_4, p_5\}$ *and* Y *a set of symptoms* $Y = \{s_1, s_2, s_3, s_4\}$ *that have to be related. Very* often it is impossible to decide that some patient definitely has some symptom or definitely has not some symptom. With fuzzy relations, this problem can be overcome by allowing partial degrees for the strength of the link between two elements. The corresponding fuzzy relation may be represented by the matrix

$$R = \begin{pmatrix} 1 & 0.2 & 0.9 & 0.8 \\ 0.2 & 0.3 & 0.2 & 0.9 \\ 0.1 & 0.9 & 1 & 0.9 \\ 0.8 & 0.2 & 0.7 & 0.9 \\ 0.9 & 0.1 & 0.1 & 0.2 \end{pmatrix}.$$

In this example, $p_2R = 0.2/s_1 + 0.3/s_2 + 0.2/s_3 + 0.9/s_4$, which gives a weighted description of the state of illness of patient p_2. $Rs_3 = 0.9/p_1 + 0.2/p_2 + 1/p_3 + 0.7/p_4 + 0.1/p_5$, which gives a weighted description of the presence of symptom s_3.

While relations are represented by matrices, the set-theoretic operations can be performed in terms of matrices.

Proposition 3.1. Let $R = (r_{ij})_{n \times m}$ and $S = (s_{ij})_{n \times m}$. Then

(1) $R \cup S = (r_{ij} \vee s_{ij})_{n \times m}$,
(2) $R \cap S = (r_{ij} \wedge s_{ij})_{n \times m}$,
(3) $R^c = (1 - r_{ij})_{n \times m}$,
(4) $R^{-1} = R^T$, where R^T stands for the transpose of R.

Proof. We prove (1) as an example.

$$(R \cup S)(x_i, y_j) = R(x_i, y_j) \vee S(x_i, y_j) = r_{ij} \vee s_{ij}$$

$(i = 1, 2, \cdots, n; j = 1, 2, \cdots, m)$, i.e.

$$R \cup S = (r_{ij} \vee s_{ij})_{n \times m}. \qquad \square$$

The following example illustrates Proposition 3.1.

Example 3.3.

$$R = \begin{pmatrix} 0.2 & 0.4 & 0.7 \\ 0.6 & 0.3 & 0.8 \\ 0.6 & 0.5 & 0.7 \end{pmatrix}, \quad S = \begin{pmatrix} 0.8 & 0.6 & 0.5 \\ 0.4 & 0.3 & 0.7 \\ 1 & 0.5 & 0.4 \end{pmatrix}.$$

Then

$$R^c = \begin{pmatrix} 0.8 & 0.6 & 0.3 \\ 0.4 & 0.7 & 0.2 \\ 0.4 & 0.5 & 0.3 \end{pmatrix}; \quad R \cap S = \begin{pmatrix} 0.2 & 0.4 & 0.5 \\ 0.4 & 0.3 & 0.7 \\ 0.6 & 0.5 & 0.4 \end{pmatrix};$$

$$R^{-1} = \begin{pmatrix} 0.2 & 0.6 & 0.6 \\ 0.4 & 0.3 & 0.5 \\ 0.7 & 0.8 & 0.7 \end{pmatrix}; \quad R_{0.6} = \begin{pmatrix} 0 & 0 & 1 \\ 1 & 0 & 1 \\ 1 & 0 & 1 \end{pmatrix};$$

$$R^c \cup S^c = (R \cap S)^c = \begin{pmatrix} 0.8 & 0.6 & 0.5 \\ 0.6 & 0.7 & 0.3 \\ 0.4 & 0.5 & 0.6 \end{pmatrix}.$$

To conclude this section, we present a general form of union, intersection and complement of fuzzy relations as in the case of fuzzy sets. Let (T, S, n) be a De Morgan triple and R, R_1, R_2 be fuzzy relations from X to Y.

(1) The complement R_n^c of R under n is defined by

$$\forall (x, y) \in X \times Y, R_n^c(x, y) = n(R(x, y)).$$

(2) The union $R_1 \cup_S R_2$ of R_1 and R_2 under S is defined by

$$\forall (x, y) \in X \times Y, (R_1 \cup_S R_2)(x, y) = S(R_1(x, y), R_2(x, y)).$$

(3) The intersection $R_1 \cap_T R_2$ of R_1 and R_2 under T is defined by

$$\forall (x, y) \in X \times Y, (R_1 \cap_T R_2)(x, y) = T(R_1(x, y), R_2(x, y)).$$

3.2 Compositions of Fuzzy Relations

As in the crisp case, there exist several different composition (product) operations including (round) composition, subcomposition, supercomposition and square composition.

3.2.1 Round Composition of Fuzzy Relations

Motivated by the characteristic function expression of the round composition of crisp relations, the round composition of two fuzzy relations is defined as follows.

Definition 3.2. *Let $R \in F(X \times Y)$, $S \in F(Y \times Z)$ and $T \in F(X \times Z)$ be three fuzzy relations. If $\forall (x, z) \in X \times Z$,*

$$T(x, z) = \bigvee_{y \in Y} (R(x, y) \wedge S(y, z)) = hgt([xR] \cap [Sz]),$$

then T is called the (round) composition of R and S, denoted by $R \circ S$.

If R is a fuzzy relation on X, we employ R^2 to denote $R \circ R$ and define R^n (n is any positive integer greater than 1) recursively by $R^n = R^{n-1} \circ R$. In the case of finite universes, the composition can be readily performed by means of matrices. To illustrate this point, let $X = \{x_1, x_2, \cdots, x_l\}$, $Y =$

$\{y_1, y_2, \cdots, y_m\}$ and $Z = \{z_1, z_2, \cdots, z_n\}$ and let $R = (r_{ij})_{l \times m}$, $S = (s_{ij})_{m \times n}$ and $T = (t_{ij})_{l \times n}$. By the definition of composition, $T = R \circ S$ means

$$T(x_i, z_k) = \bigvee_{y_j \in Y} (R(x_i, y_j) \wedge S(y_j, z_k))$$

or equivalently,

$$t_{ik} = \bigvee_{j=1}^{m} (r_{ij} \wedge s_{jk})(i = 1, 2, \cdots, l;\ k = 1, 2, \cdots, n).$$

For example, if $R = \begin{pmatrix} 0.3 & 0.7 & 0.2 \\ 1 & 0 & 0.9 \end{pmatrix}$ and $S = \begin{pmatrix} 0.8 & 0.3 \\ 0.1 & 0 \\ 0.5 & 0.6 \end{pmatrix}$, then

$R \circ S = $
$\begin{pmatrix} (0.3 \wedge 0.8) \vee (0.7 \wedge 0.1) \vee (0.2 \wedge 0.5) & (0.3 \wedge 0.3) \vee (0.7 \wedge 0) \vee (0.2 \wedge 0.6) \\ (1 \wedge 0.8) \vee (0 \wedge 0.1) \vee (0.9 \wedge 0.5) & (1 \wedge 0.3) \vee (0 \wedge 0) \vee (0.9 \wedge 0.6) \end{pmatrix}$
$= \begin{pmatrix} 0.3 & 0.3 \\ 0.8 & 0.6 \end{pmatrix}$

Proposition 3.2. *The composition of fuzzy relations fulfills the following properties provided that the involved compositions are possible to perform.*

(1) $(R \circ S) \circ T = R \circ (S \circ T)$;
(2) $R \subseteq S$ implies that $R \circ T \subseteq S \circ T$ and $T' \circ R \subseteq T' \circ S$, especially $R \subseteq S$ implies $R^n \subseteq S^n$ for any positive integer n;
(3) $(R \circ S)^{-1} = S^{-1} \circ R^{-1}$;
(4) $(R \cup S) \circ T = (R \circ T) \cup (S \circ T)$ and $T' \circ (R \cup S) = (T' \circ R) \cup (T' \circ S)$;
(5) $(R \circ S)_\alpha = R_{\dot\alpha} \circ S_{\dot\alpha}$, and if the involved universes are finite, then $(R \circ S)_\alpha = R_\alpha \circ S_\alpha$.

Proof. (1) Let $R \in F(X \times Y_1)$, $S \in F(Y_1 \times Y_2)$, $T \in F(Y_2 \times Z)$. Then $\forall (x, z) \in X \times Z$,

$$[(R \circ S) \circ T](x, z) = \bigvee_{y_2 \in Y_2} [(R \circ S)(x, y_2) \wedge T(y_2, z)]$$
$$= \bigvee_{y_2 \in Y_2} \bigvee_{y_1 \in Y_1} (R(x, y_1) \wedge S(y_1, y_2) \wedge T(y_2, z))$$
$$= \bigvee_{y_1 \in Y_1} \bigvee_{y_2 \in Y_2} (R(x, y_1) \wedge S(y_1, y_2) \wedge T(y_2, z)).$$

In the same way,

$$[R \circ (S \circ T)](x, z) = \bigvee_{y_1 \in Y_1} \bigvee_{y_2 \in Y_2} (R(x, y_1) \wedge S(y_1, y_2) \wedge T(y_2, z)).$$

Hence the equality (1) is valid.

3.2 Compositions of Fuzzy Relations

The proof of (2) and (3) is straightforward and omitted here.
(4) Let $R, S \in F(X \times Y)$ and $T \in F(Y \times Z)$. For $(x, z) \in X \times Z$,

$$
\begin{aligned}
{[(R \cup S) \circ T](x, z)} &= \bigvee_{y \in Y} [(R \cup S)(x, y) \wedge T(y, z)] \\
&= \bigvee_{y \in Y} [(R(x, y) \vee S(x, y)) \wedge T(y, z)] \\
&= \bigvee_{y \in Y} [(R(x, y) \wedge T(y, z)) \vee (S(x, y) \wedge T(y, z))] \\
&= [\bigvee_{y \in Y} (R(x, y) \wedge T(y, z))] \vee [\bigvee_{y \in Y} (S(x, y) \wedge T(y, z))] \\
&= (R \circ T)(x, z) \vee (S \circ T)(x, z) \\
&= [(R \circ T) \cup (S \circ T)](x, z).
\end{aligned}
$$

Similarly for the proof of the other equality.
(5) Let $R \in F(X \times Y)$ and $S \in F(Y \times Z)$. Then

$$
\begin{aligned}
(x, z) \in (R \circ S)_\alpha &\Leftrightarrow (R \circ S)(x, z) > \alpha \\
&\Leftrightarrow \bigvee_{y \in Y} (R(x, y) \wedge S(y, z)) > \alpha \\
&\Leftrightarrow \exists y \in Y, R(x, y) \wedge S(y, z) > \alpha \\
&\Leftrightarrow \exists y \in Y, R(x, y) > \alpha \text{ and } S(y, z) > \alpha \\
&\Leftrightarrow \exists y \in Y, (x, y) \in R_\alpha \text{ and } (y, z) \in S_\alpha \\
&\Leftrightarrow (x, z) \in R_\alpha \circ S_\alpha.
\end{aligned}
$$

The proof of the other equality is similar under the condition that the involved universes are finite. □

Remark 3.1

(1) It follows from Proposition 3.2(1) that $R^l \circ R^m = R^{l+m}$, $(R^l)^m = R^{lm}$ hold for any $R \in F(X \times X)$ and positive integers l, m.
(2) The equality $(R \cap S) \circ T = (R \circ T) \cap (S \circ T)$ does not necessarily hold as can be seen from the following example.
Let $R = \begin{pmatrix} 1 & 0 \\ 1 & 1 \end{pmatrix}$ $S = \begin{pmatrix} 0 & 1 \\ 1 & 1 \end{pmatrix}$ $T = \begin{pmatrix} 1 & 1 \\ 1 & 1 \end{pmatrix}$.
Then
$(R \cap S) \circ T = \begin{pmatrix} 0 & 0 \\ 1 & 1 \end{pmatrix}$, $(R \circ T) \cap (S \circ T) = \begin{pmatrix} 1 & 1 \\ 1 & 1 \end{pmatrix}$.
Generally, it can be verified that $(R \cap S) \circ T \subseteq (R \circ T) \cap (S \circ T)$.
(3) In Proposition 3.2(3), the equalities are still valid when the union is taken over an arbitrary index set I, i.e.

$$\left(\bigcup_{i\in I} R_i\right) \circ S = \bigcup_{i\in I}(R_i \circ S), \quad S \circ \left(\bigcup_{i\in I} R_i\right) = \bigcup_{i\in I}(S \circ R_i).$$

Finally, we point out the following:

(1) An extension of composition is the so-called T-composition with T a general t-norm. Let $R_1 \in F(X \times Y)$ and $R_2 \in F(Y \times Z)$. If $\forall (x,z) \in X \times Z$, $R(x,z) = \bigvee_{y \in Y} T(R_1(x,y), R_2(y,z))$, then R is called the T-composition of R_1 and R_2, denoted $R_1 \circ_T R_2$. Clearly, the round composition is a special T-composition in the case $T = T_{\min}$.

(2) There exists the dual concept of composition called a dual composition which is defined as follows [114]. Let $R_1 \in F(X \times Y)$ and $R_2 \in F(Y \times Z)$. If $\forall (x,z) \in X \times Z$, $R(x,z) = \bigwedge_{y \in Y}(R_1(x,y) \vee R_2(y,z))$, then R is called the dual composition of R_1 and R_2, denoted $R_1 * R_2$. Similarly, if (T, S, n) is a De Morgan triple, the dual S-composition of T-composition, denoted $*_S$, is defined by:

$$\forall (x,z) \in X \times Z, (R_1 *_S R_2)(x,z) = \bigwedge_{y \in Y}(S(R_1(x,y), R_2(y,z))).$$

3.2.2 Subcomposition, Supercomposition and Square Composition of Fuzzy Relations

Definition 3.3. Let I be a fuzzy implication, $R_1 \in F(X \times Y)$ and $R_2 \in F(Y \times Z)$. The subcomposition $R_1 \triangleleft_I R_2$ of R_1 and R_2 is defined by

$$\forall (x,z) \in X \times Z, (R_1 \triangleleft_I R_2)(x,z) = \bigwedge_{y \in Y} I(R_1(x,y), R_2(y,z)).$$

The supercomposition $R_1 \triangleright_I R_2$ of R_1 and R_2 is defined by

$$\forall (x,z) \in X \times Z, (R_1 \triangleright_I R_2)(x,z) = \bigwedge_{y \in Y} I(R_2(y,z), R_1(x,y)).$$

Proposition 3.3. *Subcomposition and supercomposition have the following properties:*

(1) $R_1 \triangleleft_I R_2 = (R_2^{-1} \triangleright_I R_1^{-1})^{-1}$ and $R_1 \triangleright_I R_2 = (R_2^{-1} \triangleleft_I R_1^{-1})^{-1}$;

(2) If I satisfies the law of contraposition, i.e. $I(x,y) = I(n(y), n(x))$, where n is a negation, then $R_1 \triangleleft_I R_2 = (R_1)_n^c \triangleright_I (R_2)_n^c$ and $R_1 \triangleright_I R_2 = (R_1)_n^c \triangleleft_I (R_2)_n^c$;

(3) If $I(x, \cdot)$ is right continuous and if I satisfies

$$\forall x, y, z \in [0,1], I(x, I(y,z)) = I(y, I(x,z)),$$

then $R_1 \triangleleft_I (R_2 \triangleright_I R_3) = (R_1 \triangleleft_I R_2) \triangleright_I R_3$.

3.2 Compositions of Fuzzy Relations

Proof. (1) Let $R_1 \in F(X \times Y)$ and $R_2 \in F(Y \times Z)$. Then $\forall (x,z) \in X \times Z$,

$$\begin{aligned}
(R_2^{-1} \triangleright_I R_1^{-1})^{-1}(x,z) &= (R_2^{-1} \triangleright_I R_1^{-1})(z,x) \\
&= \bigwedge_{y \in Y} I(R_1^{-1}(y,x), R_2^{-1}(z,y)) \\
&= \bigwedge_{y \in Y} I(R_1(x,y), R_2(y,z)) \\
&= (R_1 \triangleleft_I R_2)(x,z).
\end{aligned}$$

(2) Let $R_1 \in F(X \times Y)$ and $R_2 \in F(Y \times Z)$. Then $\forall (x,z) \in X \times Z$,

$$\begin{aligned}
[(R_1)_n^c \triangleright_I (R_2)_n^c](x,z) &= \bigwedge_{y \in Y} I((R_2)_n^c(y,z), (R_1)_n^c(x,y)) \\
&= \bigwedge_{y \in Y} I(n(R_2(y,z)), n(R_1(x,y))) \\
&= \bigwedge_{y \in Y} I(R_1(x,y), R_2(y,z)) \\
&= (R_1 \triangleleft_I R_2)(x,z).
\end{aligned}$$

(3) Let $R_1 \in F(X \times Y_1)$, $R_2 \in F(Y_1 \times Y_2)$ and $R_3 \in F(Y_2 \times Z)$. Then $\forall (x,z) \in X \times Z$,

$$\begin{aligned}
&[R_1 \triangleleft_I (R_2 \triangleright_I R_3)](x,z) \\
&= \bigwedge_{y_1 \in Y_1} I(R_1(x,y_1), (R_2 \triangleright_I R_3)(y_1,z)) \\
&= \bigwedge_{y_1 \in Y_1} I(R_1(x,y_1), \bigwedge_{y_2 \in Y_2} I(R_3(y_2,z), R_2(y_1,y_2))) \\
&= \bigwedge_{y_1 \in Y_1} \bigwedge_{y_2 \in Y_2} I(R_1(x,y_1), I(R_3(y_2,z), R_2(y_1,y_2))) \\
&= \bigwedge_{y_2 \in Y_2} \bigwedge_{y_1 \in Y_1} I(R_3(y_2,z), I(R_1(x,y_1), R_2(y_1,y_2))) \\
&= \bigwedge_{y_2 \in Y_2} I(R_3(y_2,z), \bigwedge_{y_1 \in Y_1} I(R_1(x,y_1), R_2(y_1,y_2))) \\
&= \bigwedge_{y_2 \in Y_2} I(R_3(y_2,z), (R_1 \triangleleft_I R_2)(x,y_2)) \\
&= ((R_1 \triangleleft_I R_2) \triangleright_I R_3)(x,z). \qquad \square
\end{aligned}$$

Definition 3.4. *Let E be a fuzzy equivalence on $[0,1]$, $R_1 \in F(X \times Y)$ and $R_2 \in F(Y \times Z)$. The square composition $R_1 \square_E R_2$ of R_1 and R_2 is defined by*

$$\forall (x,z) \in X \times Z, (R_1 \square_E R_2)(x,z) = \bigwedge_{y \in Y} E(R_1(x,y), R_2(y,z)).$$

Unlike the other compositions, square compositions are symmetric, i.e.

$$R_1 \square_E R_2 = R_2 \square_E R_1.$$

By Proposition 2.7, there exists a fuzzy implication I such that $\forall x \in [0,1], I(x,x) = 1$ and $\forall x,y \in [0,1], E(x,y) = \min(I(x,y), I(y,x))$. Hence,

$$(R_1 \square_E R_2)(x,z) = \bigwedge_{y \in Y} E(R_1(x,y), R_2(y,z))$$
$$= \bigwedge_{y \in Y} \min(I(R_1(x,y), R_2(y,z)), I(R_2(y,z), R_1(x,y)))$$
$$= \min(\bigwedge_{y \in Y} I(R_1(x,y), R_2(y,z)), \bigwedge_{y \in Y} I(R_2(y,z), R_1(x,y)))$$
$$= \min((R_1 \triangleleft_I R_2)(x,z), (R_1 \triangleright_I R_2)(x,z))$$
$$= [(R_1 \triangleleft_I R_2) \cap (R_1 \triangleright_I R_2)](x,z).$$

Consequently, we come to the following conclusion.

Proposition 3.4. *For every square composition $R_1 \square_E R_2$, there exists a fuzzy implication I such that $\forall x \in [0,1], I(x,x) = 1$ and $R_1 \square_E R_2 = (R_1 \triangleleft_I R_2) \cap (R_1 \triangleright_I R_2)$.*

3.3 Fuzzy Equivalence Relations

In this section, we fuzzify the concept of an equivalence relation and investigate its properties in detail. Through this and the next section, we assume that R is a fuzzy relation on X.

Definition 3.5. *If $R(x,x) = 1$ for all $x \in X$, then R is called a reflexive (fuzzy) relation.*

If X is finite and $R = (r_{ij})_{n \times n}$, reflexivity implies that $r_{ii} = 1$ ($i = 1, 2, \cdots, n$) and vice versa. As a result, we can observe the numbers on the principal diagonal of R to judge whether R is reflexive or not.

Proposition 3.5. *R is reflexive iff $\forall \alpha \in [0,1]$, R_α is reflexive.*

Proof. If R is reflexive, then $\forall \alpha \in [0,1]$, $R(x,x) = 1 \geq \alpha$. Hence $(x,x) \in R_\alpha$, viz. R_α is reflexive. Conversely, assume that $\forall \alpha \in [0,1]$, R_α is reflexive. Particularly, R_1 is reflexive. Hence $\forall x \in X$, $(x,x) \in R_1$, or $R(x,x) = 1$. □

It follows from the proof of Proposition 3.5 that R is reflexive iff R_1 (1-cut relation of R) is reflexive.

Definition 3.6. *If $\forall x, y \in X$, $R(x,y) = R(y,x)$, then R is called a symmetric (fuzzy) relation.*

Obviously, R is symmetric iff $R = R^{-1}$. We know that $R^{-1} = R^T$ in the case of finite universes. Hence R is a symmetric relation iff R as a matrix is symmetric in this case.

Proposition 3.6. *R is symmetric iff $\forall \alpha \in [0,1]$, R_α is a symmetric relation.*

Proof. If R is symmetric and $(x,y) \in R_\alpha$, then $R(y,x) = R(x,y) \geq \alpha$. Hence $(y,x) \in R_\alpha$, which proves the symmetry of R_α. Conversely, assume that

3.3 Fuzzy Equivalence Relations

$\forall \alpha \in [0,1]$, R_α is symmetric. For any $x, y \in X$, take $\alpha = R(x,y)$. Then $(x,y) \in R_\alpha$ and hence $(y,x) \in R_\alpha$ due to the symmetry of R_α. Therefore $R(y,x) \geq \alpha = R(x,y)$. A similar reasoning can lead to $R(x,y) \geq R(y,x)$. Combining the two inequalities yields $R(x,y) = R(y,x)$. □

Definition 3.7. *If $R \supseteq R^2$, then R is said to be a transitive (fuzzy) relation.*

Proposition 3.7. *R is transitive iff*
$$\forall x, y, z \in X, R(x,z) \geq R(x,y) \wedge R(y,z).$$

Proof. R is transitive $\Leftrightarrow R \supseteq R^2 \Leftrightarrow \forall x, z \in X, R(x,z) \geq R^2(x,z) \Leftrightarrow \forall x, y, z \in X, R(x,z) \geq R(x,y) \wedge R(y,z)$. □

If $X = \{x_1, x_2, \cdots, x_n\}$ is finite and $R = (r_{ij})_{n \times n}$, then R is transitive iff $R(x_i, x_k) \geq R(x_i, x_j) \wedge R(x_j, x_k)$, i.e. $r_{ik} \geq r_{ij} \wedge r_{jk} (i, j, k = 1, 2, \cdots, n)$.

Proposition 3.8. *R is transitive iff $\forall \alpha \in [0,1]$, R_α is transitive.*

Proof. Firstly, assume that R is transitive and $(x,y) \in R_\alpha$, $(y,z) \in R_\alpha$ for any fixed α in $[0,1]$. It suffices to show $(x,z) \in R_\alpha$ in order to prove the transitivity of R_α. From $(x,y) \in R_\alpha$, $(y,z) \in R_\alpha$ it follows that $R(x,y) \geq \alpha$ and $R(y,z) \geq \alpha$. Applying Proposition 3.7, we obtain $R(x,z) \geq \alpha$, namely $(x,z) \in R_\alpha$.

Conversely, assume that $\forall \alpha \in [0,1]$, R_α is transitive. We prove $\forall x, y, z \in X$, $R(x,z) \geq R(x,y) \wedge R(y,z)$. By letting $\alpha = R(x,y) \wedge R(y,z)$, we have $(x,y) \in R_\alpha$ and $(y,z) \in R_\alpha$. Hence $R(x,z) \geq \alpha = R(x,y) \wedge R(y,z)$ since R_α is transitive. Now the transitivity of R follows from Proposition 3.7. □

Remark 3.2. *A general form of fuzzy transitivity is the so-called (fuzzy) T-transitivity with T a t-norm. R is fuzzy T-transitive iff $R \supseteq R \circ_T R$, or equivalently, $\forall x, y, z \in X$, $R(x,z) \geq T(R(x,y), R(y,z))$. For instance, R is T_L-transitive iff $\forall x, y, z \in X$, $R(x,z) \geq R(x,y) + R(y,z) - 1$. Clearly, fuzzy T-transitivity is just the fuzzy transitivity when $T = T_{\min}$.*

Definition 3.8. *If R is reflexive, symmetric and transitive, then R is called a fuzzy equivalence relation.*

Proposition 3.9. *R is a fuzzy equivalence relation iff $\forall \alpha \in [0,1]$, R_α is an equivalence relation.*

Proof. It is a direct consequence of Propositions 3.5, 3.6 and 3.8. □

We know that a crisp equivalence relation determines a partition of X. So every R_α determines a partition of X if R is a fuzzy equivalence relation. For example, let R be a fuzzy relation on $X = \{x_1, x_2, x_3, x_4, x_5\}$, defined by

$$R = \begin{pmatrix} 1 & 0.4 & 0.8 & 0.5 & 0.5 \\ 0.4 & 1 & 0.4 & 0.4 & 0.4 \\ 0.8 & 0.4 & 1 & 0.5 & 0.5 \\ 0.5 & 0.4 & 0.5 & 1 & 0.6 \\ 0.5 & 0.4 & 0.5 & 0.6 & 1 \end{pmatrix}$$

Apparently, R is a reflexive, symmetric fuzzy relation. In addition, it is easily checked that $R^2 = R$, and thus R is a fuzzy equivalence relation.

For $0.8 < \alpha \leq 1$, $R_\alpha = \{(x_1, x_1), (x_2, x_2), (x_3, x_3), (x_4, x_4), (x_5, x_5)\}$, the partition of X determined by R_α is

$$\{x_1\}, \{x_2\}, \{x_3\}, \{x_4\}, \{x_5\}.$$

Similarly, for $0.6 < \alpha \leq 0.8$, the partition of X determined by R_α is

$$\{x_1, x_3\}, \{x_2\}, \{x_4\}, \{x_5\}.$$

For $0.5 < \alpha \leq 0.6$, the partition of X determined by R_α is

$$\{x_1, x_3\}, \{x_2\}, \{x_4, x_5\}.$$

For $0.4 < \alpha \leq 0.5$, the partition of X determined by R_α is

$$\{x_1, x_3, x_4, x_5\}, \{x_2\}.$$

If $\alpha \leq 0.4$, the elements in X cannot be partitioned by R_α.

Clearly, the partition determined by R_α becomes increasingly refined as the α increases.

Finally, we briefly introduce the notion of a fuzzy equivalence class.

Definition 3.9. *Let R be a fuzzy equivalence relation on X. A fuzzy set $[a]_R$ ($a \in X$) defined by:*

$$\forall x \in X, [a]_R(x) = R(a, x)$$

is called the fuzzy equivalence class of a by R. The set $X/R = \{[a]_R | a \in X\}$ of all fuzzy equivalence classes is called the fuzzy quotient set of X by R.

Example 3.4. Let $X = \{a, b, c\}$ and R be a fuzzy equivalence relation on X defined by

$$R = \begin{pmatrix} 1 & 1 & 0.7 \\ 1 & 1 & 0.7 \\ 0.7 & 0.7 & 1 \end{pmatrix}.$$

Then

$[a]_R = 1/a + 1/b + 0.7/c;$
$[b]_R = 1/a + 1/b + 0.7/c;$
$[c]_R = 0.7/a + 0.7/b + 1/c.$

The fuzzy quotient set of X by R is $X/R = \{[a]_R, [c]_R\}$.

We know that $[a]_R = [b]_R$ iff aRb in the crisp case. The following is a fuzzy counterpart of this result.

Proposition 3.10. *If R is a fuzzy equivalence relation, then*

$$[a]_R = [b]_R \text{ iff } R(a,b) = 1.$$

Proof. If $[a]_R = [b]_R$, then $R(a,b) = [a]_R(b) = [b]_R(b) = R(b,b) = 1$.
If $R(a,b) = 1$, then $\forall x \in X$,

$$[a]_R(x) = R(a,x) \geq R(a,b) \wedge R(b,x) = R(b,x) = [b]_R(x),$$

since R is transitive. Similarly, we have $[b]_R(x) \geq [a]_R(x)$. Consequently, $[a]_R = [b]_R$. □

Unlike the crisp case, the intersection of two distinct fuzzy equivalence classes may be not empty. For instance, $[a]_R \cap [c]_R = 0.7/a + 0.7/b + 0.7/c$, which is a non-empty set in Example 3.4. We have the following weaker result instead.

Proposition 3.11. *If $[a]_R \neq [b]_R$, then $\mathrm{hgt}([a]_R \cap [b]_R) < 1$.*

Proof. If $[a]_R \neq [b]_R$ and $\mathrm{hgt}([a]_R \cap [b]_R) = 1$, then due to the transitivity of R, we have

$$\begin{aligned}
R(a,b) &\geq \bigvee_{x \in X} (R(a,x) \wedge R(x,b)) \\
&= \bigvee_{x \in X} ([a]_R(x) \wedge [b]_R(x)) \\
&= \mathrm{hgt}([a]_R \cap [b]_R) = 1,
\end{aligned}$$

which contradicts Proposition 3.10.

3.4 Closures

In practical applications, we frequently use various kinds of fuzzy relations which satisfy some properties. For example, reflexivity, symmetry and transitivity are required in order to classify (partition) objects reasonably based on a fuzzy relation. However, the relation obtained from a practical problem may fail to satisfy these properties (especially transitivity) so that we have to modify them. Certainly, it makes no sense if the original relation is changed too much in the process of modification. A theoretically acceptable method is to use the so-called closure possessing the required properties to replace the original relation. In this section, we firstly introduce the general concept of a closure and then concentrate on the investigation into the transitive closure of a fuzzy relation.

3.4.1 The Concept of a Closure

Let P be a property which a fuzzy relation may satisfy or fail to satisfy. The P-closure of R is defined as the least relation (in the sense of inclusion) containing R and having the property P. More precisely, we have the following definition.

Definition 3.10. *Let R be a fuzzy relation on X. If there exists a fuzzy relation R' on X such that*

(1) $R' \supseteq R$,
(2) R' has the property P and
(3) $\forall S \in F(X \times X)$, $R \subseteq S$ and S has the property P imply $R' \subseteq S$,

then R' is called the (fuzzy) P-closure of R.

If both R' and R'' are P-closures of R, then $R' \subseteq R''$ and $R'' \subseteq R'$ by the definition of a closure, and hence $R' = R''$. In other words, the P-closure of a fuzzy relation R is unique if it exists. In this case, we denote the P-closure of R by $P(R)$. It follows from the definition that R has the property P iff $R = P(R)$.

Remark 3.3. *A fuzzy relation may have no closure for some properties. Let us take the asymmetry of a fuzzy relation to illustrate this point. A fuzzy relation R on X is called asymmetric if $\forall x, y \in X$, $R(x, y) = 0$ or $R(y, x) = 0$. Now define a fuzzy relation R on $X = \{x_1, x_2\}$ by*

$$R = \begin{pmatrix} 0 & 0.3 \\ 0.2 & 0.1 \end{pmatrix}.$$

Then R does not have an asymmetric closure.

Now we present a necessary and sufficient condition for the existence of a P-closure.

Theorem 3.1. *A P-closure exists for all $R \in F(X \times X)$ iff*

(1) The entire relation $E = X \times X$ has the property P and
(2) The intersection of every non-empty family of fuzzy relations, each of which has the property P, has P as well.

Proof. Suppose that a P-closure exists for all $R \in F(X \times X)$. Particularly, $P(E)$ exists. By the definition of a closure, $E \subseteq P(E)$, whence $E = P(E)$ since $P(E) \subseteq E$ trivially holds. As a result, E possesses P. To prove (2), let S be the intersection of a non-empty family of fuzzy relations $\mathcal{S}$, each of which has the property P. Considering that $S \subseteq R$ and R has P for every $R \in \mathcal{S}$, we have $P(S) \subseteq R$. Hence, $P(S) \subseteq S$, whence $P(S) = S$. It follows that S has P.

Conversely, suppose that (1) and (2) are satisfied. Let R be an arbitrary relation on X. Let $\mathcal{S}$ be the family of all relations, each of which contains

R and has P. This family is not empty since E is one of them by (1). Then the intersection R' of this family has P by (2). In addition, $R' \supseteq R$ and $\forall S \in F(X \times X)$, $R \subseteq S$ and S has P imply $R' \subseteq S$. Therefore, R' is the P-closure of R. □

By applying Theorem 3.1 and Exercise 7 in Section 3.4, it can be checked that the reflexive closure, symmetric closure and transitive closure of every fuzzy relation do exist.

3.4.2 The Transitive Closure of a Fuzzy Relation

This subsection will be devoted to the investigation of a special closure – the fuzzy transitive closure. We firstly introduce a general formula for computing the transitive closure of a fuzzy relation.

Proposition 3.12. *The transitive closure of a fuzzy relation R on X is $R' = \bigcup_{k=1}^{\infty} R^k$.*

Proof. We prove that $R' = \bigcup_{k=1}^{\infty} R^k$ is the least transitive fuzzy relation containing R. Firstly, $R' \supseteq R$ is obvious.

Secondly, $R' \circ R' = \bigcup_{k=1}^{\infty} R^k \circ \bigcup_{l=1}^{\infty} R^l = \bigcup_{k=1}^{\infty}\bigcup_{l=1}^{\infty} (R^k \circ R^l) = \bigcup_{m=2}^{\infty} R^k \subseteq R'$, i.e. $R' \supseteq (R')^2$. Hence R' is transitive.

Finally, assume that S is a fuzzy relation on X such that $S \supseteq R$ and S is transitive. Then $S \supseteq S^2 \supseteq R^2$. By mathematical induction, we can prove that $S \supseteq R^k$ holds for any positive integer k. Consequently, $S \supseteq R'$, which completes the proof. □

In the sequel, we use $t(R)$ to denote the transitive closure of R and $|X|$ the cardinality of X. If X is finite, the formula for computing the fuzzy transitive closure of R can be simplified.

Proposition 3.13. *If $|X| = n$, then $t(R) = \bigcup_{k=1}^{n} R^k$.*

Proof. By mathematical induction, it can be proved that for an arbitrary positive integer m,

$$R^m(x,z) = \bigvee_{y_1 \in X} \bigvee_{y_2 \in X} \cdots \bigvee_{y_{m-1} \in X} R(x,y_1) \wedge R(y_1,y_2) \wedge \cdots \wedge R(y_{m-1},z).$$

Hence

$$R^{n+1}(x,z) = \bigvee_{y_1 \in X} \bigvee_{y_2 \in X} \cdots \bigvee_{y_n \in X} R(x,y_1) \wedge R(y_1,y_2) \wedge \cdots \wedge R(y_n,z);$$

whence there exist $y_1, y_2, \cdots, y_n \in X$ such that

$$R^{n+1}(x, z) = R(x, y_1) \wedge R(y_1, y_2) \wedge \cdots \wedge R(y_n, z).$$

Since $x(= y_0), y_1, y_2, \cdots, y_n$ are $n+1$ elements in X, there must exist y_i and y_j $(i < j)$ such that $y_i = y_j$. Then

$$R^{n+1}(x, z) \leq R(x, y_1) \wedge \cdots \wedge R(y_{i-1}, y_i) \wedge R(y_j, y_{j+1}) \wedge \cdots \wedge R(y_n, z),$$

whence

$$R^{n+1}(x, z) \leq R^{n+1-(j-i)}(x, z) \leq \bigvee_{i=1}^{n} R^i(x, z) = (\bigcup_{i=1}^{n} R^i)(x, z),$$

and thus $R^{n+1} \subseteq \bigcup_{i=1}^{n} R^i$. Now

$$R^{n+2} = R^{n+1} \circ R \subseteq (\bigcup_{i=1}^{n} R^i) \circ R = \bigcup_{i=2}^{n+1} R^i = (\bigcup_{i=2}^{n} R^i) \cup R^{n+1} \subseteq \bigcup_{i=1}^{n} R^i.$$

In a recursive manner, we can verify that $R^m \subseteq \bigcup_{i=1}^{n} R^i$ for any integer $m \geq n$, which implies that

$$t(R) = \bigcup_{i=1}^{\infty} R^i = \bigcup_{i=1}^{n} R^i. \qquad \square$$

3.5 Fuzzy Tolerance Relations

An equivalence relation has desired behavior and can be applied conveniently. However it is often difficult for a fuzzy relation to meet the transitivity requirement. In this section, we introduce a class of fuzzy relations without transitivity.

Definition 3.11. *If a fuzzy relation R on X is reflexive and symmetric, then R is called a fuzzy tolerance relation.*

Remark 3.4. *In the literature, a reflexive and symmetric fuzzy relation is also called a compatibility relation [83] or a similarity relation [98].*

In the case of a finite universe, a tolerance relation corresponds to a symmetric matrix with all elements on the principle diagonal line being 1's.

Proposition 3.14. *If R is a fuzzy tolerance relation, then R^k is a fuzzy tolerance relation for every positive integer k.*

3.5 Fuzzy Tolerance Relations

Proof. When $k = 1$, $R^k = R$ is a fuzzy tolerance relation. Assume that R^m is a fuzzy tolerance relation. We consider the fuzzy relation R^{m+1}.

For all $x \in X$,

$$R^{m+1}(x,x) = \bigvee_{y \in X} (R^m(x,y) \wedge R(y,x)) \geq R^m(x,x) \wedge R(x,x) = 1,$$

which implies that $R^{m+1}(x,x) = 1$, i.e. R^{m+1} is reflexive.

In addition, $\forall x, y \in X$,

$$R^{m+1}(x,y) = \bigvee_{z \in X} (R^m(x,z) \wedge R(z,y))$$
$$= \bigvee_{z \in X} (R^m(z,x) \wedge R(y,z))$$
$$= R^{m+1}(y,x),$$

i.e. R^{m+1} is symmetric. Therefore R^{m+1} is a fuzzy tolerance relation. The desired result follows from the principle of mathematical induction. □

Proposition 3.15. *If $|X| = n$ and R is a fuzzy tolerance relation on X, then $t(R) = R^n$.*

Proof. By Proposition 3.13, $t(R) = \bigcup_{k=1}^{n} R^k$. It suffices to show that $R^m \subseteq R^{m+1}$ holds for any positive integer m. Indeed, $\forall x, y \in X$,

$$R^2(x,y) = \bigvee_{z \in X} (R(x,z) \wedge R(z,y))$$
$$\geq R(x,y) \wedge R(y,y) = R(x,y)$$

due to the reflexivity of R. It means that $R \subseteq R^2$. The inclusion $R^m \subseteq R^{m+1}$ may be recursively obtained. □

For example, for the fuzzy tolerance relation $R = \begin{pmatrix} 1 & 0.6 & 0.5 \\ 0.6 & 1 & 0.3 \\ 0.5 & 0.3 & 1 \end{pmatrix}$, we have

$$R^2 = \begin{pmatrix} 1 & 0.6 & 0.5 \\ 0.6 & 1 & 0.5 \\ 0.5 & 0.5 & 1 \end{pmatrix} \text{ and } R^3 = R^2. \text{ Hence}$$

$$t(R) = \begin{pmatrix} 1 & 0.6 & 0.5 \\ 0.6 & 1 & 0.5 \\ 0.5 & 0.5 & 1 \end{pmatrix}.$$

Corollary 3.1. *If $|X| = n$ and R is a fuzzy tolerance relation on X, then $R^m = R^n$ as $m > n$.*

Proof. By the proof of Proposition 3.15, $R^n \subseteq R^m$. On the other hand, $R^m \subseteq \bigcup_{i=1}^{\infty} R^i = t(R) = R^n$. So they are identical. □

Corollary 3.2. *If $|X| = n$ and R is a fuzzy tolerance relation on X, then $t(R)$ is a fuzzy equivalence relation.*

Proof. By Proposition 3.14, $t(R) = R^n$ is reflexive and symmetric. In addition, $t(R)$ as the transitive closure of R is transitive. Hence $t(R)$ is a fuzzy equivalence relation. □

To acquire the transitive closure of a fuzzy tolerance relation R, we should calculate R, R^2, R^3, …, R^n so that n times composition operations need to be performed. However, the process can be accelerated due to Corollary 3.1. Firstly, calculate R^2. If $R = R^2$, then R itself is transitive and $t(R) = R$; otherwise, calculate $R^4 = R^2 \circ R^2$. If $R^4 = R^2$, then $t(R) = R^2$; otherwise, repeat this process until $R^m = R^{2m}$ and the transitive closure is R^m. The process may also stop and $t(R) = R^m$ if we have obtained R^m for some $m \geq n$.

Example 3.5. *If*

$$R = \begin{pmatrix} 1 & 0.1 & 0.8 & 0.5 & 0.3 \\ 0.1 & 1 & 0.1 & 0.2 & 0.4 \\ 0.8 & 0.1 & 1 & 0.5 & 0.5 \\ 0.5 & 0.2 & 0.5 & 1 & 0.6 \\ 0.3 & 0.4 & 0.5 & 0.6 & 1 \end{pmatrix},$$

then R is a fuzzy tolerance relation. Simple calculations yield

$$R^2 = \begin{pmatrix} 1 & 0.3 & 0.8 & 0.5 & 0.5 \\ 0.3 & 1 & 0.4 & 0.4 & 0.4 \\ 0.8 & 0.4 & 1 & 0.5 & 0.5 \\ 0.5 & 0.4 & 0.5 & 1 & 0.6 \\ 0.5 & 0.4 & 0.5 & 0.6 & 1 \end{pmatrix}$$

and

$$R^4 = \begin{pmatrix} 1 & 0.4 & 0.8 & 0.5 & 0.5 \\ 0.4 & 1 & 0.4 & 0.4 & 0.4 \\ 0.8 & 0.4 & 1 & 0.5 & 0.5 \\ 0.5 & 0.4 & 0.5 & 1 & 0.6 \\ 0.5 & 0.4 & 0.5 & 0.6 & 1 \end{pmatrix}.$$

Further calculations lead to $R^8 = R^4$, which indicates $t(R) = R^4$.

3.6 Other Special Fuzzy Relations

In some previous sections, fuzzy equivalence and similarity relations are introduced. Undoubtedly, these relations have some important applications,

3.6 Other Special Fuzzy Relations

e.g. in the classification of objects (see Section 2.8.1). However, they are far from enough considering the wide applications of fuzzy relations. In decision analysis, one often needs to compare alternatives. Fuzzy binary relations are suitable for representing pairwise comparisons between objects, which leads to the theory of fuzzy preference modeling. In this theory, many transitivity-related concepts are involved in order to conceptualize various preference structures. These concepts can also be found in the study of fuzzy choice functions or ranking fuzzy quantities. In view of those, this section is devoted to a brief introduction of some of these notions, including mainly negative S-transitivity, T-S-semitransitivity, T-S-Ferrers property, where T and S are a t-norm and a t-conorm respectively.

3.6.1 Crisp Negative Transitivity, Semitransitivity, and Ferrers Property

Let R be a crisp binary relation on a given set X. Some notions associated with R are defined as follows.

Definition 3.12

(1) R is called *irreflexive* if R satisfies that: $\forall x \in X$, $xR^c x$.
(2) R is called *complete* if R satisfies that: $\forall x, y \in X, x \neq y$, xRy or yRx.
(3) R is called *antisymmetric* if R satisfies that: $\forall x, y \in X, x \neq y$, xRy implies $yR^c x$.
(4) R is called *negatively transitive* if R satisfies that: $(\forall x, y, z \in X)$, $(xRz \Rightarrow xRy$ or $yRz)$.
(5) R is called *semitransitive* if R satisfies that: $\forall x_1, x_2, y_1, y_2 \in X$, $x_1 R y_1$ and $y_1 R y_2$ imply $x_1 R x_2$ or $x_2 R y_2$.
(6) R is called a *Ferrers* (or *semiorder*) relation if R satisfies that:

$$\forall x_1, x_2, y_1, y_2 \in X, x_1 R y_1 \text{ and } x_2 R y_2 \text{ imply } x_1 R y_2 \text{ or } x_2 R y_1.$$

Example 3.6. Let $f : X \to \mathbb{R}$ and $q > 0$ a real number. For $x_1, x_2 \in \mathbb{R}$, define R by: $x_1 R x_2$ if $f(x_1) > f(x_2) + q$. Then it can be verified that R is a semitransitive relation.

Example 3.7. Let I denote the set of all finite open intervals on $\mathbb{R}$. For $]a_1, b_1[,]a_2, b_2[\in I$, define $<$ by: $]a_1, b_1[<]a_2, b_2[$ if $b_1 < a_2$. It can be checked that $<$ is a Ferrers relation.

It is worth noting that properties of R and R^c are closely related.

Proposition 3.16

(1) R is reflexive iff R^c is irreflexive;
(2) R is complete iff R^c is antisymmetric;
(3) R is negatively transitive iff R^c is transitive;

(4) R is semitransitive iff R^c is semitransitive;
(5) R is a Ferrers relation iff R^c is a Ferrers relation.

Proof. The proof of these equivalencies are straightforward and omitted here.
□

Remark 3.5. *If two properties P and P' of relations satisfy that: every relation R has the property P iff the complement R^c of R has the property P', then the properties P and P' are called two dual properties. Therefore, reflexivity and irreflexivity, completeness and antisymmetry, negative transitivity and transitivity are dual properties respectively while semitransitivity and Ferrers property are both self-dual.*

There exist the following relationships among the above defined properties.

Proposition 3.17

(1) *If R is complete and transitive, then R is a negatively transitive, semitransitive and Ferrers relation.*
(2) *If R is antisymmetric and negatively transitive, then R is a transitive, semitransitive and Ferrers relation.*
(3) *If R is irreflexive and semitransitive (or Ferrers), then R is transitive.*
(4) *If R is reflexive and semitransitive (or Ferrers), then R is negatively transitive.*
(5) *If R is transitive and negatively transitive, then R is a semitransitive and Ferrers relation.*

Proof. As examples, we prove (3) and (5).

(3) Firstly, suppose that R is irreflexive and semitransitive. If aRb and bRc, then semitransitivity implies that aRa or aRc. By the irreflexivity of R, aRc is obtained. Hence, R is transitive.

Next, suppose that R is an irreflexive and Ferrers relation. If aRb and bRc, then we have aRc or bRb. By the irreflexivity of R, aRc is obtained, which means that R is transitive.

(5) If aRb and bRc, then aRc by the transitivity of R. For any $d \in X$, it follows from the negative transitivity of R that aRd or dRc, which implies that R is semitransitive.

In order to prove the Ferrers property, let aRb and cRd. Then aRc or cRb by the negative transitivity of R. In the case of cRb, the Ferrers property holds. If aRc, together with cRd, we have aRd by the transitivity of R. The Ferrers property also holds.
□

It follows immediately from Proposition 3.17 that:

(1) If R is a complete, irreflexive and semitransitive (Ferrers) relation, then R is a Ferrers (semitransitive) relation;
(2) If R is a reflexive, antisymmetric, and semitransitive (Ferrers) relation then R is a Ferrers (semitransitive) relation.

3.6.2 Negative S-Transitivity, T-S-Semitransitivity, and T-S-Ferrers Property

In this subsection, we extend the concepts in the last subsection to the fuzzy case and present some corresponding results.

Definition 3.13. *Let R be a fuzzy relation on X and (T, S, n) a De Morgan triple.*

(1) If R satisfies that: $\forall a, b \in X, a \neq b, T(R(a,b), R(b,a)) = 0$, then R is called T-antisymmetric;
(2) If R satisfies that: $\forall a, b \in X, a \neq b, S(R(a,b), R(b,a)) = 1$, then R is called S-complete;
(3) R is called irreflexive if $\forall a \in X, R(a,a) = 0$;
(4) If R satisfies that: $\forall a, b, c \in X, S(R(a,b), R(b,c)) \geq R(a,c)$, then R is called negatively S-transitive;
(5) If R satisfies that:

$$\forall a, b, c, d \in X, S(R(a,b), R(b,c)) \geq T(R(a,d), R(d,c)),$$

then R is called T-S-semitransitive;
(6) If R satisfies that:

$$\forall a, b, c, d \in X, S(R(a,b), R(c,d)) \geq T(R(a,d), R(c,b)),$$

then R is called a T-S-Ferrers relation.

Now we investigate two special cases. Firstly, consider the case $T(x,y) = \min(x,y)$ or $S(x,y) = \max(x,y)$. For convenience, we shall drop the prefix min and max. For instance, max-complete, min-antisymmetric, min-max-semitransitive are simply called complete, antisymmetric and semitransitive respectively. In this case, all the notions involved in Definition 3.13 are the so-called cutworthy ones, i.e. R has the property P iff R_α has the property P for all $\alpha \in [0, 1]$. More specifically, we have the following results.

Proposition 3.18. *Let R be a fuzzy relation on X.*

(1) R is antisymmetric iff R_α is antisymmetric for all $\alpha \in [0, 1]$;
(2) R is complete iff R_α is complete for all $\alpha \in [0, 1]$;
(3) R is negatively transitive iff R_α is negatively transitive for all $\alpha \in [0, 1]$;
(4) R is semitransitive iff R_α is semitransitive for all $\alpha \in [0, 1]$;
(5) R is a Ferrers relation iff R_α is a Ferrers relation for all $\alpha \in [0, 1]$.

Proof. We prove (5) as an example.

Firstly, suppose that R is a Ferrers relation. For any fixed $\alpha \in [0, 1]$, let $aR_\alpha b$ and $cR_\alpha d$. Then $R(a,b) \geq \alpha$ and $R(c,d) \geq \alpha$. By the Ferrers property of R, we have

$$\max\{R(a,d), R(c,b)\} \geq \min\{R(a,b), R(c,d)\} \geq \alpha,$$

whence $R(a,d) \geq \alpha$ or $R(c,b) \geq \alpha$, or equivalently, $aR_\alpha d$ or $cR_\alpha b$, which indicates that R_α is a Ferrers relation.

Conversely, suppose that R_α is a Ferrers relation for any $\alpha \in [0,1]$.

For any $a,b,c,d \in X$, let $\alpha = \min\{R(a,b), R(c,d)\}$. Then $R(a,b) \geq \alpha$ and $R(c,d) \geq \alpha$, i.e. $aR_\alpha b$ and $cR_\alpha d$. By the Ferrers property of R_α, we have $aR_\alpha d$ or $cR_\alpha b$, In other words, $R(a,d) \geq \alpha$ or $R(c,b) \geq \alpha$, or equivalently, $\max\{R(a,d), R(c,b)\} \geq \alpha$. Hence

$$\max\{R(a,d), R(c,b)\} \geq \min\{R(a,b), R(c,d)\},$$

and thus R is a Ferrers relation. $\square$

The second special case is for a strong De Morgan triple $(T_L^\varphi, S_L^\varphi, N_\varphi)$. In this case, it can be checked that:

(1) R is S-complete iff

$$\forall a,b \in X, a \neq b, \varphi(R(a,b)) + \varphi(R(b,a)) \geq 1.$$

(2) R is T-antisymmetric iff

$$\forall a,b \in X, a \neq b, \varphi(R(a,b)) + \varphi(R(b,a)) \leq 1.$$

(3) R is negatively S-transitive iff

$$\forall a,b,c \in X, \varphi(R(a,b)) \leq \varphi(R(a,c)) + \varphi(R(c,b)).$$

(4) R is T-S-semitransitive iff

$$\forall a,b,c,d \in X, \varphi(R(a,c)) + \varphi(R(c,b)) - 1 \leq \varphi(R(a,d)) + \varphi(R(d,b)).$$

(5) R is T-S-Ferrers relation iff

$$\forall a,b,c,d \in X, \varphi(R(a,b)) + \varphi(R(c,d)) - 1 \leq \varphi(R(a,d)) + \varphi(R(c,b)).$$

We firstly present a proposition, which is a counterpart of Proposition 3.16.

Proposition 3.19. *Let (T,S,n) be a De Morgan triple with n a strong negation. Then*

(1) R is T-antisymmetric iff R_n^c is S-complete;
(2) R is T-transitive iff R_n^c is negatively S-transitive;
(3) R is T-S-semitransitive iff R_n^c is T-S-semitransitive;
(4) R is a T-S-Ferrers relation iff R_n^c is a T-S-Ferrers relation.

Proof. (1) R is T-antisymmetric

$$\Leftrightarrow \forall a,b \in X, a \neq b, T(R(a,b), R(b,a)) = 0$$
$$\Leftrightarrow n(T(n(R_n^c(a,b)), n(R_n^c(b,a)))) = n(0) = 1 (a \neq b)$$

3.6 Other Special Fuzzy Relations

$$\Leftrightarrow S(R_n^c(a,b), R_n^c(b,a)) = 1 (a \neq b)$$
$$\Leftrightarrow R_n^c \text{ is } S-\text{complete}.$$

(2) R is T-transitive $\Leftrightarrow \forall a, b, c \in X, T(R(a,b), R(b,c)) \leq R(a,c)$
$$\Leftrightarrow n(T(n(R_n^c(a,b)), n(R_n^c(b,c)))) \geq n(R(a,c)),$$
$$\Leftrightarrow S(R_n^c(a,b), R_n^c(b,c)) \geq R_n^c(a,c)$$
$$\Leftrightarrow R_n^c \text{ is negatively } S\text{-transitive}.$$

The proof of (3) and (4) is similar to that of (2) and omitted. □

Now we start to deal with the relationships between negative S-transitivity, T-transitivity, T-S-semitransitivity and T-S-Ferrers property. Firstly, we point out that a part of Proposition 3.17(5) is still true in the fuzzy case, which is stated as follows.

Proposition 3.20. *If R is T-transitive and negatively S-transitive, then R is a T-S-semitransitive relation.*

Proof. By the T-transitivity and negative S-transitivity of R, for any $a, b, c, d \in X$, we have

$$T(R(a,b), R(b,c)) \leq R(a,c) \leq S(R(a,d), R(d,c)).$$

Hence R is a T-S-semitransitive relation. □

Any other result in Proposition 3.17 is not necessarily true even though T is continuous and n is a strong negation. The following example illustrates that S-completeness and T-transitivity cannot guarantee negative S-transitivity.

Example 3.8. Let $X = \{a, b, c\}$ and R be defined as

$$R = \begin{pmatrix} 1 & 0.7 & 0.3 \\ 0.7 & 1 & 0.4 \\ 0.3 & 0.2 & 1 \end{pmatrix}.$$

Let $n(x) = 1 - x$, $T(x,y) = \varphi^{-1}(max(\varphi(x) + \varphi(y) - 1, 0))$ with $\varphi(x) = x^2$. Then

$$S(x,y) = n(T(n(x), n(y))) = (\varphi \circ n)^{-1}(max((\varphi \circ n)(x) + (\varphi \circ n)(y) - 1, 0)).$$

It is easily checked that: $\forall x, y, z \in X$,

$$(\varphi \circ n)(R(x,y)) + (\varphi \circ n)(R(y,x)) \leq 1$$

and

$$\varphi(R(x,y)) + 1 \geq \varphi(R(x,z)) + \varphi(R(z,y)),$$

which indicate that R is S-complete and T-transitive. However,

$$(\varphi \circ n)(R(a,c)) + (\varphi \circ n)(R(c,b)) > (\varphi \circ n)(R(a,b)) + 1,$$

i.e. R is not negatively S-transitive.

However, Proposition 3.17 is still true in some special cases.

Proposition 3.21. Let R be a fuzzy relation on X and (T, S, n) a De Morgan triple.

(1) If R is complete and T-transitive, then R is a negatively S-transitive, T-S-semitransitive and T-S-Ferrers relation.
(2) If R is antisymmetric and negatively S-transitive, then R is a T-transitive, T-S-semitransitive and T-S-Ferrers relation.
(3) If R is an irreflexive and T-S-semitransitive (T-S-Ferrers) relation, then R is T-transitive.
(4) If R is a reflexive and T-S-semitransitive (T-S-Ferrers) relation, then R is negatively $S-$transitive.

Proof. (1) Firstly, we prove that

$$\forall a, b, c \in X, S(R(a,b), R(b,c)) \geq R(a,c).$$

If $b = c$, the inequality is obvious. So we assume $b \neq c$. In this case, the inequality is again obvious if $R(b,c) = 1$. Otherwise, $R(c,b) = 1$ by the completeness of R. Hence

$$S(R(a,b), R(b,c)) \geq R(a,b) \geq T(R(a,c), R(c,b)) = R(a,c).$$

Therefore, R is negatively S-transitive.

Secondly, we prove that

$$\forall a, b, c, d \in X, S(R(a,b), R(b,c)) \geq T(R(a,d), R(d,c)).$$

If $a = b$,

$$S(R(a,b), R(b,c)) = S(R(a,a), R(a,c)) \geq R(a,c) \geq T(R(a,d), R(d,c)).$$

The inequality holds. Now suppose $a \neq b$. When $R(a,b) = 1$, the inequality apparently holds. Otherwise, $R(b,a) = 1$ by the completeness of R.

$$\begin{aligned}S(R(a,b), R(b,c)) &\geq R(b,c) \\ &\geq T(R(b,a), R(a,c)) = R(a,c) \\ &\geq T(R(a,d), R(d,c)),\end{aligned}$$

i.e., R is T-S-semitransitive.

The proof of T-S-Ferrers property is similar to that of T-S-semitransitivity.

3.6 Other Special Fuzzy Relations

(2) We only prove T-S-semitransitivity. The remaining proofs are similar. It suffices to show that

$$\forall a,b,c,d \in X, S(R(a,b), R(b,c)) \geq T(R(a,d), R(d,c)).$$

If $a = d$, then

$$T(R(a,d), R(d,c)) = T(R(a,a), R(a,c)) \leq R(a,c) \leq S(R(a,b), R(b,c)),$$

which is the desired inequality. We assume $a \neq d$. When $R(a,d) = 0$, the inequality is obvious. Otherwise, $R(d,a) = 0$ by the antisymmetry of R.

$$\begin{aligned} T(R(a,d), R(d,c)) &\leq R(d,c) \\ &\leq S(R(d,a), R(a,c)) = R(a,c) \\ &\leq S(R(a,b), R(b,c)). \end{aligned}$$

(3) If R is T-S-semitransitive, then for any $a, b, c \in X$, due to the irreflexivity of R,

$$T(R(a,b), R(b,c)) \leq S(R(a,a), R(a,c)) = R(a,c),$$

i.e. R is T-transitive.

If R is a T-S-Ferrers relation, then for any $a, b, c \in X$, due to the irreflexivity of R,

$$T(R(a,b), R(b,c)) \leq S(R(a,c), R(b,b)) = R(a,c),$$

i.e. R is T-transitive.

(4) Similar to the proof of (3). □

It follows immediately from Proposition 3.21 that:

(1) If R is a complete, irreflexive and T-S-semitransitive (T-S-Ferrers) relation, then R is a T-S-Ferrers (T-S-semitransitive) relation.

(2) If R is a reflexive, antisymmetric, and T-S-semitransitive (T-S-Ferrers) fuzzy relation then R is a T-S-Ferrers (T-S-semitransitivity) fuzzy relation.

The condition completeness or antisymmetry of R is very strong, which will be weakened in the coming discussion. Certainly, we have to pay a price which is a restriction on the De Morgan triple.

Proposition 3.22. *Let (T, S, n) be a De Morgan triple satisfying the following condition:*

$$(\mathcal{C}) \quad \forall x, y, z \in [0,1], T(S(x,y), z) \leq S(x, T(y,z)).$$

(1) If R is S-complete and T-transitive, then R is a negatively S-transitive, T-S-semitransitive and T-S-Ferrers relation.

(2) If R is T-antisymmetric and negatively S-transitive, then R is a T-transitive, T-S-semitransitive and T-S-Ferrers relation.

Proof. (1) Firstly, we prove the negative S-transitivity, i.e.

$$\forall a,b,c \in X, R(a,c) \leq S(R(a,b), R(b,c)).$$

If $b = c$, it is obvious.

If $b \neq c$, then $S(R(b,c), R(c,b)) = 1$ by the S-completeness of R. Hence, by the T-transitivity and the condition $(\mathcal{C})$,

$$\begin{aligned}S(R(a,b), R(b,c)) &\geq S(T(R(a,c), R(c,b)), R(b,c)) \\ &\geq T(S(R(b,c), R(c,b)), R(a,c)) = R(a,c),\end{aligned}$$

i.e. R satisfies the negative S-transitivity.

Now the T-S-semitransitivity follows from Proposition 3.20.

What is left is the proof of Ferrers property, i.e.

$$\forall a,b,c,d \in X, T(R(a,b), R(c,d)) \leq S(R(a,d), R(c,b)).$$

Indeed, by the negative S-transitivity, condition $(\mathcal{C})$ and T-transitivity, we obtain successively,

$$\begin{aligned}T(R(a,b), R(c,d)) &\leq T(S(R(a,d), R(d,b)), R(c,d)) \\ &\leq S(R(a,d), T(R(c,d), R(d,b))) \\ &\leq S(R(a,d), R(c,b)),\end{aligned}$$

which completes the proof of (1).

(2) We firstly prove the T-transitivity of R, i.e.

$$\forall a,b,c \in X, T(R(a,b), R(b,c)) \leq R(a,c).$$

If $b = c$, the inequality is clearly valid.

If $b \neq c$, $T(R(b,c), R(c,b)) = 0$ by the antisymmetry of R. Hence, by the negative S-transitivity and condition $(\mathcal{C})$,

$$\begin{aligned}T(R(a,b), R(b,c)) &\leq T(S(R(a,c), R(c,b)), R(b,c)) \\ &\leq S(R(a,c), T(R(c,b), R(b,c))) \\ &= R(a,c).\end{aligned}$$

T-S-semitransitivity follows from Proposition 3.20 and the proof of T-S-Ferrers property is the same as that of the T-S-Ferrers property in (1). $\square$

In the presence of condition $(\mathcal{C})$, combining Proposition 3.22 and 3.21 may yield some new results. For example,

3.6 Other Special Fuzzy Relations

(i) If R is an S-complete, irreflexive and T-S-semitransitive (T-S-Ferrers) relation, then R is a T-S-Ferrers (T-S-semitransitive) relation;

(ii) If R is a T-antisymmetric, reflexive, T-S-semitransitive (T-S-Ferrers) relation then R is a T-S-Ferrers (T-S-semitransitive) relation, etc. In addition, it follows from the proof of Proposition 3.22(1) that:

If condition ($\mathcal{C}$) is satisfied, then T-transitivity and negative S-transitivity imply T-S-Ferrers property.

Finally, we would like to say more about condition ($\mathcal{C}$). This condition is based on the crisp set equality:

$$(A \cup B) \cap C \subseteq A \cup (B \cap C).$$

It is easily checked that all of (T_L, S_L), (T_π, S_π) and $(T_{\min}, S_{\max})$ satisfy the condition.

3.6.3 Consistency, Weak Transitivity and Acyclicity

Let R be a fuzzy relation on X. In decision making, X usually stands for the set of alternatives and $R(x,y)$ ($x, y \in X$) represents the degree of "x is at least as good as y". Then the comparison of alternative x and y becomes that of $R(x,y)$ and $R(y,x)$. In this situation, the following notions are helpful.

Definition 3.14. *Let R be a fuzzy relation on X.*

(1) If $(\forall a, b, c \in X)$, $(R(a,b) \geq R(b,a)$ and $R(b,c) \geq R(c,b))$ imply $R(a,c) \geq R(c,a))$, then R is called consistent;

(2) If $(\forall a, b, c \in X)$, $(R(a,b) > R(b,a)$ and $R(b,c) > R(c,b))$ imply $R(a,c) > R(c,a))$, then R is called weakly transitive;

(3) If there are no $a_1, a_2, \ldots, a_m \in X$ such that

$$R(a_1, a_2) > R(a_2, a_1), R(a_2, a_3) > R(a_3, a_2), \ldots, R(a_{m-1}, a_m) >$$

$R(a_m, a_{m-1})$ and $R(a_m, a_1) > R(a_1, a_m)$, then R is called acyclic.

Given a fuzzy relation R on X, define the crisp relations

$$R' = \{(a,b) | R(a,b) \geq R(b,a)\}$$

and

$$R'' = \{(a,b) | R(a,b) > R(b,a)\}.$$

Clearly, R is consistent iff R' is transitive, and R is weakly transitive iff R'' is transitive.

It follows from the definition that weak transitivity implies acyclicity.

Proposition 3.23. *If R is transitive, then R is weakly transitive.*

Proof. We prove it by contradiction. Suppose that $R(a,b) > R(b,a)$, $R(b,c) > R(c,b)$ and $R(a,c) \leq R(c,a)$. By transitivity,

$$R(a,c) \geq \min(R(a,b), R(b,c)).$$

If $R(a,b) \leq R(b,a)$, then $R(c,a) \geq R(a,c) \geq R(a,b)$, and thus,

$$R(b,a) \geq \min(R(b,c), R(c,a)) \geq R(a,b),$$

a contradiction.

If $R(a,b) > R(b,a)$, then $R(a,c) > R(b,c)$ and thus

$$R(b,c) > R(c,b) \geq \min(R(c,a), R(a,b)) \geq \min(R(a,c), R(b,c)) = R(b,c),$$

a contradiction. □

Proposition 3.24. *A fuzzy relation R on X is consistent iff the following statements are true:*

(1) R is weakly transitive;
(2) $(\forall a,b,c \in X)$, $(R(a,b) = R(b,a)$ and $R(b,c) = R(c,b)$ imply $R(a,c) = R(c,a))$.

Proof. Let R be consistent. Firstly, we prove the weak transitivity of R.

If $R(a,b) > R(b,a)$, $R(b,c) > R(c,b)$ and $R(a,c) \leq R(c,a)$, then $R(c,a) \geq R(a,c)$, $R(a,b) > R(b,a)$, and $R(b,c) > R(c,b)$, which violates the consistency.

Conversely, suppose that (1) and (2) hold.

If $R(a,b) \geq R(b,a)$, $R(b,c) \geq R(c,b)$ and $R(a,c) < R(c,a)$.

(1) If $R(a,b) > R(b,a)$, $R(b,c) > R(c,b)$, then $R(a,c) < R(c,a)$, which contradicts (1).
(2) If $R(a,b) = R(b,a)$, $R(b,c) > R(c,b)$, then $R(a,c) < R(c,a)$, which contradicts (1).
(3) $R(a,b) > R(b,a)$, $R(b,c) = R(c,b)$, then $R(a,c) < R(c,a)$, which contradicts (1).
(4) $R(a,b) = R(b,a)$, $R(b,c) = R(c,b)$, then $R(a,c) < R(c,a)$, which contradicts (2). □

Therefore, consistency implies weak transitivity.

Generally speaking, consistency and transitivity are independent which can be illustrated by the following example.

Example 3.9. *Let $X = \{a,b,c\}$, R_1 and R_2 be fuzzy relations on X defined as:*

$$R_1 = \begin{pmatrix} 1 & 0.5 & 0.7 \\ 0.5 & 1 & 0.5 \\ 0.8 & 0.5 & 1 \end{pmatrix} \quad R_2 = \begin{pmatrix} 1 & 0.9 & 0.6 \\ 0.7 & 1 & 0.8 \\ 0.5 & 0.7 & 1 \end{pmatrix}.$$

Then it can be checked that R_1 is transitive and R_2 is consistent. However, $R_1(a,b) = R_1(b,a)$, $R_1(b,c) = R_1(c,b)$ and $R_1(a,c) < R_1(c,a)$, which imply

3.6 Other Special Fuzzy Relations

that R_1 is not consistent. In addition, $R_2(a,c) < \min\{R_2(a,b), R_2(b,c)\}$, which implies R_2 is not transitive.

To conclude this section, we present an important result in fuzzy decision making as an application of acyclicity.

Proposition 3.25. *Let R be a fuzzy relation on a finite set X. For $B \in P(X)$, define*

$$R(B) = \{a | a \in B, \forall b \in B, R(a,b) \geq R(b,a)\}.$$

Then R is acyclic iff $R(B) \neq \emptyset$ holds for all $B \in P(X)$.

Proof. Firstly, assume that $R(B) \neq \emptyset$ holds for all $B \in P(X)$. We prove that R is acyclic by contradiction.

Suppose that there exist $a_1, a_2, \ldots, a_m \in X$ such that

$$R(a_1, a_2) > R(a_2, a_1), R(a_2, a_3) > R(a_3, a_2), \ldots, R(a_{m-1}, a_m) > R(a_m, a_{m-1})$$

and $R(a_m, a_1) > R(a_1, a_m)$. By letting $B = \{a_1, a_2, \ldots, a_m\}$, we have $R(B) = \emptyset$ which is in contradiction to our assumption.

Conversely, assume that R is acyclic. We prove that $R(B) \neq \emptyset$ holds for all $B \in P(X)$. Suppose, to the contrary, there exists $B \in P(X)$ such that $R(B) = \emptyset$. Take an arbitrary $a_1 \in B$. Then there must exist $a_2 \in B$ such that $R(a_2, a_1) > R(a_1, a_2)$ due to $a_1 \notin R(B)$. Similarly, there must exist $a_3 \in B$ such that $R(a_3, a_2) > R(a_2, a_3)$ due to $a_2 \notin R(B)$. Continuing in this way, we have $a_n \in B$ satisfying $R(a_{n+1}, a_n) > R(a_n, a_{n+1})$ for any positive integer n. From the finiteness of X, we infer that there must exist $a_i \in B$ $(1 \leq i \leq n)$ such that $a_i = a_n$ when $n > |X|$. Now we have $a_i, a_{i+1}, \ldots, a_{n+1} \in B \subseteq X$ such that

$$R(a_{i+1}, a_i) > R(a_i, a_{i+1}), \ldots, R(a_n, a_{n-1}) > R(a_n, a_{n-1})$$

and $R(a_{n+1}, a_n) > R(a_n, a_{n+1})$, i.e.,

$$R(a_{i+1}, a_i) > R(a_i, a_{i+1}), \ldots, R(a_n, a_{n-1}) > R(a_n, a_{n-1})$$

and $R(a_{n+1}, a_i) > R(a_i, a_{n+1})$, which is in contradiction to acyclicity. □

Remark 3.6. *Relations and their properties are extensively employed in decision making analysis. As we know, one of the central tasks in a decision making problem is to rank all the involved alternatives or choose some "good" alternatives. To fulfill this task, a decision maker has to make comparisons among alternatives. An extensively used approach to comparing alternatives is pairwise comparison, which can be represented by a relation called preference. This approach has been well-developed preference modeling theory [119]. In this theory, relations are employed to model various comparing results between alternatives including strict preference, indifference, incomparability and large preference.*

Some properties of relations, e.g. reflexivity, completeness, transitivity, semitransitivity, Ferrers property are among the most important concepts for modeling particular preference structures such as weak order, partial order, interval order, semitransitive order etc. In 1982, Orlovsky [109] used a fuzzy relation to express the preference degrees between alternatives and developed the concept of fuzzy preference. From then, the choice function based on a fuzzy preference is investigated in detail by some authors [14, 18, 85, 103, 120, 131] in which acyclicity, weak transitivity and consistency are frequently used properties. Meanwhile, preference structures are also fuzzified with the fuzzification of preference. For the literature on fuzzy preference structures, we recommend the references [20, 21, 34, 35, 54, 56, 93]. As in the crisp case, the fuzzy counterparts of transitivity, semitransitivity and Ferrers property play an important role in the investigation of particular fuzzy preference structures.

3.7 Fuzzy Relation Equations

A fuzzy relation equation is an equality containing unknown fuzzy relations. For example, given fuzzy relations R and S, $X \cup R = S$, $X^2 \cap R = X \circ S$ with X an unknown fuzzy relation are fuzzy relation equations. In this section, we concentrate on the investigation into the type of equations $X \circ R = S$, where R and S are given relations and X is the unknown fuzzy relation. Our aim is to find X such that $X \circ R = S$ is satisfied. For this purpose, we firstly introduce some notions concerning the equation in the following.

Definition 3.15. *If a fuzzy relation X satisfies the equation $X \circ R = S$, then X is called a solution. An equation which has some solutions is called consistent. For a consistent equation, if a solution $\overline{X}$ is such that for any other solution X, $X \subseteq \overline{X}$, then $\overline{X}$ is called the maximal solution.*

Specifically, let $R \in F(V \times W)$ and $S \in F(U \times W)$. Then $X \in F(U \times V)$. The equation $X \circ R = S$ is written as

$$\forall (u,w) \in U \times W, \bigvee_{v \in V} (X(u,v) \wedge R(v,w)) = S(u,w).$$

Take the notation

$$\overline{X}(u,v) = \bigwedge_w \{S(u,w) | S(u,w) < R(v,w)\}.$$

We make the convention that $\bigvee_{i \in \emptyset} \alpha_i = 0$ and $\bigwedge_{i \in \emptyset} \alpha_i = 1$ henceforth.

Theorem 3.2. *The equation $X \circ R = S$ is consistent iff $\overline{X} \circ R = S$.*

Proof. The sufficiency is obvious.

3.7 Fuzzy Relation Equations

Necessity. Assume that X is an arbitrary solution of $X \circ R = S$. Then

$$\forall (u,w) \in U \times W, \bigvee_{v \in V} (X(u,v) \wedge R(v,w)) = S(u,w).$$

Hence, the inequality $X(u,v) \wedge R(v,w) \leq S(u,w)$ $((\forall u,v,w) \in U \times V \times W)$ holds. For any fixed u,v, $X(u,v) \leq S(u,w)$ if $S(u,w) < R(v,w)$. As a result,

$$X(u,v) \leq \bigwedge_{w} \{S(u,w)|S(u,w) < R(v,w)\} = \overline{X}(u,v).$$

If the inequality $S(u,w) \geq R(v,w)$ holds for any w in W, then $\overline{X}(u,v) = 1$ since $\{S(u,w)|S(u,w) < R(v,w)\} = \emptyset$. We still have $X(u,v) \leq \overline{X}(u,v)$, and thus $X \subseteq \overline{X}$.

On the other hand, if $S(u,w) < R(v,w)$, then $\overline{X}(u,v) \leq S(u,w)$ by the definition of $\overline{X}$. Otherwise, $S(u,w) \geq R(v,w)$. In either case, we have $\overline{X}(u,v) \wedge R(v,w) \leq S(u,w)$ $(\forall (u,v,w) \in U \times V \times W)$. It follows that

$$\forall (u,w) \in U \times W, \bigvee_{v \in V} (\overline{X}(u,v) \wedge R(v,w)) \leq S(u,w),$$

i.e. $\overline{X} \circ R \subseteq S$. Now we have $S = X \circ R \subseteq \overline{X} \circ R \subseteq S$, which implies the desired result $\overline{X} \circ R = S$. $\square$

Remark 3.7. *It follows from the proof of the theorem that*
(1) If the equation $X \circ R = S$ is consistent, then $\overline{X}$ defined by

$$\overline{X}(u,v) = \bigwedge_{w} \{S(u,w)|S(u,w) < R(v,w)\}$$

is the maximal solution;
(2) The inclusion $\overline{X} \circ R \subseteq S$ holds no matter whether the equation is consistent or not.

In the sequel, we shall restrict our attention to fuzzy relation equations on finite universes. Let us start with the equations $X \circ R = S$, where $X = (x_1, x_2, \cdots, x_n)$, $S = (s_1, s_2, \cdots, s_m)$ and $R = (r_{ij})_{n \times m}$. In this case, the equations may be explicitly written as

$$\bigvee_{k=1}^{n} (x_k \wedge r_{kj}) = s_j (j = 1, 2, \cdots, m).$$

By Remark 3.7(1), the maximum solution would be $\overline{X} = \{\bar{x}_1, \bar{x}_2, \cdots, \bar{x}_n\}$, where $\bar{x}_k = \bigwedge_j \{s_j | s_j < r_{kj}\}$ provided that the equation is consistent. Take the notation

$$G = \{(k_1, k_2, \cdots, k_m) | \bar{x}_{k_j} \wedge r_{k_j j} = s_j, j = 1, 2 \cdots, m\}.$$

If $G \neq \emptyset$, for $g = (k_1, k_2, \cdots, k_m) \in G$ and $k = 1, 2, \cdots, n$, let

$$x_{g_k} = \bigvee_j \{s_j | k_j = k\}.$$

Write $X_g = (x_{g_1}, x_{g_2}, \cdots, x_{g_n})$.

Theorem 3.3. *The equation $X \circ R = S$ is consistent iff $G \neq \emptyset$. Under the condition $G \neq \emptyset$, we can find g in G such that $X_g \subseteq X \subseteq \overline{X}$ for every solution X.*

Proof. We need to show that

(i) If the equation is consistent, then $G \neq \emptyset$;
(ii) If $G \neq \emptyset$, then the equation is consistent;
(iii) For every solution X, an element g in G can be found such that $X_g \subseteq X \subseteq \overline{X}$.

Firstly, we prove (i).

Since the equation is consistent, the maximal solution is $\overline{X}$. Thus

$$\bigvee_{k=1}^n (\overline{x}_k \wedge r_{kj}) = s_j (j = 1, 2, \cdots, m).$$

Hence, there exist $k_1, k_2, \cdots, k_m$ such that $\overline{x}_{k_j} \wedge r_{k_j j} = s_j (j = 1, 2 \cdots, m)$, which implies $(k_1, k_2, \cdots, k_m) \in G$, i.e. $G \neq \emptyset$.

Next we prove (iii).

Assume that $X = (x_1, x_2, \cdots, x_n)$ is a solution. Similar to the proof of (i), there exist $k_1, k_2, \cdots, k_m$ such that $x_{k_j} \wedge r_{k_j j} = s_j (j = 1, 2 \cdots, m)$. Hence

$$s_j = x_{k_j} \wedge r_{k_j j} \leq \overline{x}_{k_j} \wedge r_{k_j j} \leq \bigvee_{k=1}^n (\overline{x}_k \wedge r_{kj}) \leq s_j.$$

As a result, $\overline{x}_{k_j} \wedge r_{k_j j} = s_j (j = 1, 2 \cdots, m)$ and thus $g = (k_1, k_2, \cdots, k_m) \in G$. Now it suffices to prove $X_g \subseteq X$ considering that $X \subseteq \overline{X}$ is trivially true.

For any fixed k ($1 \leq k \leq n$), if there exist some $k'_j s$, $k_j = k$, then $x_k = x_{k_j} \geq s_j$ and thus $x_k \geq \bigvee_j \{s_j | k_j = k\} = x_{g_k}$.

If there is no k_j equal to k, then $x_{g_k} = 0 \leq x_k$.

Therefore, $x_{g_k} \leq x_k$ holds for all $k = 1, 2, \cdots, n$. Consequently, $X_g \subseteq X$.

Finally, we prove (ii).

Putting $g = (k_1, k_2, \cdots, k_m) \in G$, we prove that $X_g = (x_{g_1}, x_{g_2}, \cdots, x_{g_n})$ is a solution.

Since

$$\bigvee_{k=1}^n (x_{g_k} \wedge r_{kj}) \geq x_{g_{k_j}} \wedge r_{k_j j} = \bigvee_i \{s_i | k_i = k_j\} \wedge r_{k_j j} \geq s_j \wedge r_{k_j j} = s_j,$$

we have $X_g \circ R \supseteq S$.

3.7 Fuzzy Relation Equations

For any fixed k $(1 \leq k \leq n)$, if there exist some $k'_j s$, $k_j = k$, then $\bar{x}_k = \bar{x}_{k_j} \geq s_j$ and $\bar{x}_k \geq \bigvee_j \{s_j | k_j = k\} = x_{g_k}$.

If there is no k_j equal to k, then $x_{g_k} = 0 \leq \bar{x}_k$.

Therefore $X_g \subseteq \overline{X}$. In summary, we have $S \subseteq X_g \circ R \subseteq \overline{X} \circ R \subseteq S$, which implies that $X_g \circ R = S$. Hence X_g is a solution. □

It can be seen from the proof of Theorem 3.3 that if $G \neq \emptyset$, then every X satisfying that $X_g \subseteq X \subseteq \overline{X}$ is a solution. Consequently, the set of solutions of the equation is

$$\mathcal{X} = \{X | X_g \subseteq X \subseteq \overline{X}, g \in G\}.$$

Example 3.10. Solve the equation

$$(x_1, x_2, x_3) \circ \begin{pmatrix} 0.3 & 0.5 & 0.2 \\ 0.2 & 0 & 0.4 \\ 0 & 0.6 & 0.1 \end{pmatrix} = (0.2, 0.4, 0.2).$$

In this example,

$$\bar{x}_1 = \bigwedge_j \{s_j | s_j < r_{1j}\} = 0.2 \wedge 0.4 = 0.2.$$

$$\bar{x}_2 = \bigwedge_j \{s_j | s_j < r_{2j}\} = 0.2.$$

$$\bar{x}_3 = \bigwedge_j \{s_j | s_j < r_{3j}\} = 0.4.$$

$\bar{x}_{k_1} \wedge r_{k_1 1} = s_1 = 0.2$ leads to $k_1 \in \{1, 2\}$. Similarly, $k_2 = 3$ and $k_3 \in \{1, 2\}$. Hence

$$G = \{(1, 3, 1), (1, 3, 2), (2, 3, 1), (2, 3, 2)\}.$$

For $g = (k_1, k_2, k_3) = (1, 3, 1)$, we have

$$x_{g_1} = \bigvee_j \{s_j | k_j = 1\} = 0.2 \vee 0.2 = 0.2.$$

$$x_{g_2} = \bigvee_j \{s_j | k_j = 2\} = 0.$$

$$x_{g_3} = \bigvee_j \{s_j | k_j = 3\} = 0.4.$$

Thus $X_g = (x_{g_1}, x_{g_2}, x_{g_3}) = (0.2, 0, 0.4)$. The set of solutions satisfying that $X_g \subseteq X \subseteq \overline{X}$ is $(0.2, [0, 0.2], 0.4)$. In the same manner, the corresponding set of solutions for $g = (2, 3, 1)$ or $g = (1, 3, 2)$ is $(0.2, 0.2, 0.4)$, which is contained in the set $(0.2, [0, 0.2], 0.4)$ and is dropped. The set of solutions for $g = (2, 3, 2)$ is $([0, 0.2], 0.2, 0.4)$. In summary, the set of solutions in the example is

$$\mathcal{X} = \{X | X_g \subseteq X \subseteq \overline{X}, g \in G\} = \{(0.2, [0, 0.2], 0.4), ([0, 0.2], 0.2, 0.4)\}.$$

The preceding procedure of solving a fuzzy relation equation can be arranged in the following table.

	x_1	x_2	x_3
$s_1 = 0.2$	0.2	0.2	
$s_2 = 0.4$			0.4
$s_3 = 0.2$	0.2	0.2	
$\overline{X}$	0.2	0.2	0.4
$(1,3,1)$	0.2	0	0.4
$(2,3,2)$	0	0.2	0.4
$(1,3,2)$	0.2	0.2	0.4
$(2,3,1)$	0.2	0.2	0.4

Based on this table, the set of solutions can be quickly acquired by comparing $\overline{X}$ to all X_g ($g \in G$), which are listed in the lower part of the table.

More generally, consider the equation $X \circ R = S$, where $X = (x_{ij})_{n \times m}$, $R = (r_{ij})_{m \times l}$ and $S = (s_{ij})_{n \times l}$. This equation is equivalent to the following system of fuzzy relation equations:

$$(x_{i1}, x_{i2}, \cdots, x_{im}) \circ R = (s_{i1}, s_{i2}, \cdots, s_{il})(i = 1, 2, \cdots, n),$$

which are n equations solvable by means of the preceding procedure.

Example 3.11. *Solve the equation*

$$\begin{pmatrix} x_{11} & x_{12} & x_{13} \\ x_{21} & x_{22} & x_{23} \end{pmatrix} \circ \begin{pmatrix} 0.3 & 0.5 & 0.2 \\ 0.2 & 0.1 & 0 \\ 0 & 0.4 & 0.3 \end{pmatrix} = \begin{pmatrix} 0 & 0.4 & 0.3 \\ 0.2 & 0.2 & 0.2 \end{pmatrix}.$$

Decompose the equation into

$$(x_{11}, x_{12}, x_{13}) \circ \begin{pmatrix} 0.3 & 0.5 & 0.2 \\ 0.2 & 0.1 & 0 \\ 0 & 0.4 & 0.3 \end{pmatrix} = (0, 0.4, 0.3)$$

and

$$(x_{21}, x_{22}, x_{23}) \circ \begin{pmatrix} 0.3 & 0.5 & 0.2 \\ 0.2 & 0.1 & 0 \\ 0 & 0.4 & 0.3 \end{pmatrix} = (0.2, 0.2, 0.2).$$

Tabling the solving process of two equations, we obtain

	x_{11}	x_{12}	x_{13}
$s_{11} = 0$	0	0	0
$s_{12} = 0.4$			0.4
$s_{13} = 0.3$			0.3
$\overline{X}_1$	0	0	1
X_{1g}	0	0	0.4

	x_{21}	x_{22}	x_{23}
$s_{21} = 0.2$	0.2	0.2	
$s_{22} = 0.2$	0.2		0.2
$s_{23} = 0.2$	0.2		0.2
$\overline{X}_2$	0.2	1	0.2
X_{2g}	0.2	0	0
	0	0.2	0.2

Based on these tables, the set of solutions is

$$\left\{ \begin{pmatrix} 0 & 0 & [0.4,1] \\ 0.2 & [0,1] & [0,0.2] \end{pmatrix}, \begin{pmatrix} 0 & 0 & [0.4,1] \\ [0,0.2] & [0.2,1] & 0.2 \end{pmatrix} \right\}$$

Finally, consider the equation $R \circ X = S$. By Proposition 3.2, it can be transformed into $X^{-1} \circ R^{-1} = S^{-1}$, which is solvable.

Remark 3.8. *The study of fuzzy relation equations was initiated by Sanchez [127] in 1976. The equations involved in this section are the so-called* sup $-$ min *composite fuzzy relation equations on the unit interval. For more general* sup $-T$ *equations with* T *a t-norm or those formulated on a lattice, the reader may refer to [17, 32, 39, 41, 128, 144]. Besides this type of equations, many researchers investigated various other types of equations such as* $s - t$ *equations[112],* inf $-\mathcal{I}$ *composite equations [11, 116] with* $\mathcal{I}$ *a fuzzy implication, fuzzy bilinear equations [140] and so on. If the solution set is empty, the study of approximate solutions may be carried out [63, 113]. For applications of fuzzy relation equations in fuzzy reasoning, fuzzy control and other areas, see [29, 40, 65, 111, 129]. In [33, 40, 42], an overview of all related research can be found.*

3.8 Some Applications of Fuzzy Relations

In this section, we shall introduce some applications of fuzzy relations.

3.8.1 Fuzzy Clustering Analysis

Objects are normally described by means of multiple indices in a real world problem. The mathematical method in which these multiple indices are employed to classify objects is called clustering analysis. Fuzzy clustering analysis refers to the mathematical method in which fuzzy set theory is applied for the purpose of classification.

Let $X = \{x_1, x_2, \ldots, x_n\}$ be the set of objects to be classified. Assume that $C_1, C_2, \ldots, C_m$ are m indices. The corresponding indices of x_i ($1 \leq i \leq n$) are $x_{i1}, x_{i2}, \ldots, x_{im}$. Now the question is how to classify X based on these given indices. As we know, this problem is attacked in Mathematical Statistics. Fuzzy relation theory provides an alternative solution to the problem which is stated in detail in the sequel.

Firstly, transform the indices $(x_{i1}, x_{i2}, \ldots, x_{im})$ and $(x_{j1}, x_{j2}, \ldots, x_{jm})$ of x_i and x_j into a single number r_{ij} which reflects the similarity degree of x_i with x_j. This process is called the scaling process. The scaling approaches in the literature are various. Here are some of them.

(1) **Scalar product approach**

$$r_{ij} = \begin{cases} 1 & i = j \\ \dfrac{1}{M}\sum_{k=1}^{m} x_{ik}x_{jk} & i \neq j \end{cases}$$

where $M = \max\limits_{i \neq j} |\sum_{k=1}^{m} x_{ik}x_{jk}|$.

(2) **Correlation coefficient approach**

$$r_{ij} = \frac{\sum_{k=1}^{m}(x_{ik} - \bar{x}_i)(x_{jk} - \bar{x}_j)}{\sqrt{\sum_{k=1}^{m}(x_{ik} - \bar{x}_i)^2}\sqrt{\sum_{k=1}^{m}(x_{jk} - \bar{x}_j)^2}},$$

where $\bar{x}_i = \dfrac{1}{m}\sum_{k=1}^{m} x_{ik}$ $(i = 1, 2, \ldots, n)$.

(3) **Max-Min approach**

$$r_{ij} = \frac{\sum_{k=1}^{m}(x_{ik} \wedge x_{jk})}{\sum_{k=1}^{m}(x_{ik} \vee x_{jk})}.$$

(4) **Exponential approach**

$$r_{ij} = e^{-\sum_{k=1}^{m}|x_{ik} - x_{jk}|}.$$

(5) **Distance approach**

$$r_{ij} = 1 - c\sum_{k=1}^{m}|x_{ik} - x_{jk}|,$$

where c is a properly chosen constant so that $r_{ij} \in [0, 1]$ $(i, j = 1, 2, \ldots n)$.

It can be seen that $r_{ij} \in [-1, 1]$ after scaling whatever approach mentioned above is applied. If $r_{ij} \in [-1, 0]$ for some i, j, then for all $i = 1, 2, \cdots, n$ and $j = 1, 2, \cdots, n$, let

$$r'_{ij} = \frac{1}{2}(1 + r_{ij})$$

or

$$r'_{ij} = \frac{r_{ij} - \min_{i \neq j} r_{ij}}{(\max_{i \neq j} - \min_{i \neq j})}$$

so that $r'_{ij} \in [0,1]$ for all $i,j = 1,2,\cdots,n$. For convenience, we assume that $r_{ij} \in [0,1]$ after scaling. In addition, it is clear that $r_{ii} = 1$, $r_{ij} = r_{ji}$ ($i,j = 1,2,\ldots,n$). Now write $R = (r_{ij})_{n\times n}$, which is a fuzzy tolerance relation. Calculate the transitive closure $t(R)$ of R and use it to classify $x_1, x_2, \ldots, x_n$ at some level α.

Example 3.12. *Every environmental cell consists of air, water, soil, and plants which are called four crucial elements in assessing the environmental quality of the cell. The content of pollutants in the four elements can reflect how much the environmental cell is polluted. For each element, there is a prescribed content value of pollutants. The more the content of pollutants in a certain element exceeds the prescribed value, the more the environmental cell is polluted in terms of the element. In the following table, the pollution data pertaining to four environmental cells are listed.*

	air	water	soil	plants
I	5	5	3	2
II	2	3	4	5
III	5	5	2	3
IV	1	5	3	1
I	2	4	5	1

The numbers in the table reflect how much the contents of pollutants go beyond the prescribed values. The bigger the number is, the more the content of pollutants exceeds the prescribed value of the element in the column. Based on these pollution data, we can classify the five environmental cells I, II, III, IV, V.

Firstly, we obtain the similarity evaluations between environmental cells by scaling the data. The distance approach is taken for this purpose, i.e.

$$r_{ij} = 1 - c \sum_{k=1}^{4} |x_{ik} - x_{jk}| \text{ with } c = 0.1.$$

For instance, $r_{12} = 1 - 0.1(|5-2| + |5-3| + |3-4| + |2-5|) = 0.1$, which means that the similarity evaluation is 0.1 between the environmental cell I and II. In a similar way, all r_{ij} ($i,j = 1,2,\ldots,n$) can be calculated and thus we obtain the matrix

$$R = \begin{pmatrix} 1 & 0.1 & 0.8 & 0.5 & 0.3 \\ 0.1 & 1 & 0.1 & 0.2 & 0.4 \\ 0.8 & 0.1 & 1 & 0.3 & 0.1 \\ 0.5 & 0.2 & 0.3 & 1 & 0.6 \\ 0.3 & 0.4 & 0.1 & 0.6 & 1 \end{pmatrix}.$$

This is the very relation in Example 3.5 and thus the classification based on the transitive closure $t(R)$ at the different levels may refer to the example.

In practice, it is not necessary to compute the transitive closure $t(R)$ in order to get the classifications at different levels if R is a fuzzy tolerance relation. This is the direct clustering method introduced in the following. We take Example 3.12 for an illustration.

For $\alpha = 1$, $r_{ii} = 1$ $(i = 1, 2, \ldots, n)$ and hence the classification is $\{x_1\}$, $\{x_2\}$, $\{x_3\}$, $\{x_4\}$, $\{x_5\}$.

For $\alpha = 0.8$, $r_{13} = 0.8$. Merging the class $\{x_1\}$ and $\{x_3\}$ at the level $\alpha = 1$, we have the classification $\{x_1, x_3\}$, $\{x_2\}$, $\{x_4\}$, $\{x_5\}$.

For $\alpha = 0.6$, $r_{45} = 0.6$. Merging the class $\{x_4\}$ and $\{x_5\}$ at the level $\alpha = 0.8$, we have the classification $\{x_1, x_3\}$, $\{x_2\}$, $\{x_4, x_5\}$.

Similarly, the classification is $\{x_1, x_3, x_4, x_5\}$, $\{x_2\}$ at the level $\alpha = 0.5$ and X can not be classified at $\alpha \leq 0.4$.

Here is another more complex example.

Example 3.13. Let $X = \{x_1, x_2, x_3, x_4, x_5, x_6, x_7\}$. $R \in F(X \times X)$ is defined in the form of a matrix

$$R = \begin{pmatrix} 1 & 0.8 & 1 & 0.2 & 0.8 & 0.5 & 0.3 \\ 0.8 & 1 & 0.4 & 0.3 & 0.7 & 0.6 & 0.3 \\ 1 & 0.4 & 1 & 0.7 & 1 & 0.6 & 0.5 \\ 0.2 & 0.3 & 0.7 & 1 & 0.5 & 0.8 & 0.6 \\ 0.8 & 0.7 & 1 & 0.5 & 1 & 0.2 & 0.7 \\ 0.5 & 0.6 & 0.6 & 0.8 & 0.2 & 1 & 0.8 \\ 0.3 & 0.3 & 0.5 & 0.6 & 0.7 & 0.8 & 1 \end{pmatrix}.$$

For the greatest value $\alpha = 1$, $r_{ii} = 1$ $(i = 1, 2, \ldots, n)$, $r_{13} = r_{35} = 1$, and thus x_1, x_3, x_5 should be in the same class. As a consequence, the classification is $\{x_1, x_3, x_5\}$, $\{x_2\}$, $\{x_4\}$, $\{x_6\}$, $\{x_7\}$.

For the second greatest value $\alpha = 0.8$, $r_{12} = r_{46} = r_{67} = 0.8$. Merging the class $\{x_1, x_3, x_5\}$ and $\{x_2\}$, $\{x_4\}$, $\{x_6\}$ and $\{x_7\}$ at the level $\alpha = 0.8$, we have the classification $\{x_1, x_2, x_3, x_5\}$, $\{x_4, x_6, x_7\}$.

For $\alpha = 0.7$, $r_{34} = 0.7$, and hence all the elements would be put together.

3.8.2 An Application to Information Retrieval

In a document retrieval system there are basically three finite sets involved: a set $U = \{u_1, u_2, \cdots, u_p\}$ of users of the system, a set $D = \{d_1, d_2, \cdots, d_q\}$ of documents constituting the system and a set $T = \{t_1, t_2, \cdots, t_r\}$ of terms describing the documents. The document retrieval system can be described by means of a fuzzy relation F from D to T, where $F(d,t)$ indicates for every pair (d,t) belonging to $D \times T$ to what degree a document deals with a term t. It should be stressed that there is a great difference between the importance of a term t in the description of a document d and the degree a document d treats of a term t. If, for example, a very short and uninteresting document deals almost exclusively with "politics" then the importance of the term "politics" for that document will be near 1. On the other hand,

3.8 Some Applications of Fuzzy Relations

the importance of that document for the term "politics" will surely be near zero. From the users' point of view the latter is preferable for, if they want documents about politics, the paper described will give them no satisfaction. For this reason, the relation F indicates the degree to which a document is relevant to a term rather than indicating the importance of the term in this document.

The following is an introduction to the work by Zenner et al. [177]. As set of documents, a selection of 20 articles with various length from papers, weeklies, magazines, all dealing to some extent with the following terms: politics, finances, economics, employment, cruise missiles, U.S.A. and Russia.

A heterogeneous group of 15 persons (consisting of professors, teachers, laborers, secretaries, housewives and students) was then asked to ascertain to which degree (between 0 and 1), in their opinion, the documents deal with the terms considered. Then several methods were developed to come up with a common document description relation. The following relation is extracted from this experiment. We restrict ourselves here to six documents $d_1, d_2, \cdots, d_6$ and five terms $t_1, t_2, \cdots, t_5$, obtaining the following document description relation F:

$$F = \begin{pmatrix} 0.8 & 0.6 & 0 & 0 & 0.6 \\ 0.6 & 0.5 & 0.6 & 0.1 & 0.1 \\ 0.2 & 0.2 & 0 & 0 & 0.4 \\ 0.3 & 0.3 & 0 & 0 & 0 \\ 0.8 & 0.8 & 0.3 & 0.1 & 0.6 \\ 0.3 & 0.4 & 0 & 0 & 0.2 \end{pmatrix}.$$

where $F(d_i, t_j)$ represents the degree to which document d_i deals with term t_j.

Using the concepts of afterset and foreset, we may deduce some interesting information from F. For example, if we are interested in the set of documents that deal with term t_1, we may consider the F-foreset of t_1:

$$Ft_1 = 0.8/d_1 + 0.6/d_2 + 0.2/d_3 + 0.3/d_4 + 0.8/d_5 + 0.3/d_6.$$

Similarly, the F-afterset of document d_3:

$$d_3F = 0.2/t_1 + 0.2/t_2 + 0.4/t_5$$

indicates the fuzzy set of terms that are treated in this document.

In addition, let us consider the meaning of FcF^{-1} and $F^{-1}cF$, where c stands for some composition. Clearly, for every choice of c, FcF^{-1} is a fuzzy relation on D and $F^{-1}cF$ is a fuzzy relation on T. We shall calculate membership degrees and interpret them in terms of information retrieval.

(1) Since $d_1F = 0.8/t_1 + 0.6/t_2 + 0.6/t_5$ and $d_3F = 0.2/t_1 + 0.2/t_2 + 0.4/t_5$, $(F \circ F^{-1})(d_1, d_3) = hgt(d_1F \cap F^{-1}d_3) = hgt(d_1F \cap d_3F) = 0.4$, which may

be interpreted as the degree in which there exists at least one term that is treated in d_1 as well as in d_3, in other words, it is a measure for the existing overlap of the information included in documents d_1 and d_3.

Similarly, $(F^{-1} \circ F)(t_1, t_3) = hgt(Ft_1 \cap Ft_3) = 0.6$, which may be interpreted as the degree to which there exists at least one document that deals with term t_1 as well as with term t_3, i.e. a measure for the overlap of Ft_1 and Ft_3.

(2) Let $I(x, y) = \max\{1 - x, y\}$. Then we have
$$(F \triangleleft_I F^{-1})(d_1, d_3) = \bigwedge_{j=1}^{5} (1 - F(d_1, t_j) \vee F^{-1}(t_j, d_3)) = \bigwedge_{j=1}^{5} (1 - F(d_1, t_j) \vee F(d_3, t_j)) = 0.2,$$
which may be interpreted as a measure for the degree to which every term treated in document d_1 is also treated in document d_3, in other words as a measure for the degree to which information included in d_1 is also included in document d_3.

Similarly, calculating $(F^{-1} \triangleleft_I F)$ yields $(F^{-1} \triangleleft_I F)(t_1, t_3) = 0.2$, which may be interpreted as a measure for the degree to which every document that deals with term t_1, also deals with t_3.

(3) Let $I(x, y) = \max\{1 - x, y\}$. Then we have
$$(F \triangleright_I F^{-1})(d_1, d_3) = \bigwedge_{j=1}^{5} (1 - F(d_3, t_j) \vee F^{-1}(t_j, d_1)) = \bigwedge_{j=1}^{5} (1 - F(d_3, t_j) \vee F(d_1, t_j)) = 0.6,$$
which may be interpreted as a measure for the degree to which every term treated in document d_3 is also treated in document d_1, in other words as a measure for the degree to which information included in d_3 is also included in document d_1.

Similarly, calculating $(F^{-1} \triangleright_I F)$ yields $(F^{-1} \triangleright_I F)(t_1, t_3) = 0.6$, which may be interpreted as a measure for the degree to which every document that deals with term t_3, also deals with t_1.

(4) Let $E(x, y) = \min\{\max\{1 - x, y\}, \max\{1 - y, x\}\}$. Then we have $(F \square_E F^{-1})(d_1, d_3) = 0.2$, which may be interpreted as a measure for the degree to which every term treated in document d_1 is also treated in document d_3 and vice versa.

(5) Let $E(x, y) = \min\{\max\{1 - x, y\}, \max\{1 - y, x\}\}$. Then we have $(F^{-1} \square_E F)(t_1, t_3) = 0.2$, which may be interpreted as a measure for the degree to which every document that deals with term t_3, also deals with t_1 and vice versa.

3.8.3 An Application to Multiple Attribute Decision Making Analysis

Multiple attribute decision making (MADM) refers to making a decision (ranking or choosing some alternatives) based on multiple, frequently conflicting, attributes. These attributes may be goals or criteria. MADM is very common in our daily life. For example, one may choose a job among several

3.8 Some Applications of Fuzzy Relations

offers by taking the attributes salary, promotion opportunity, work location etc. into account. When we buy a car, we may consider the attributes like price, color, safety, gas mileage etc. Generally, assume that $A_1, A_2, \ldots, A_n$ are n alternatives and $C_1, C_2, \ldots, C_m$ are m attributes. An MADM problem can be represented in the form of a matrix:

$$R = \begin{pmatrix} r_{11} & r_{12} & \cdots & r_{1m} \\ r_{21} & r_{22} & \cdots & r_{2m} \\ \cdots & \cdots & & \\ r_{n1} & r_{n2} & \cdots & r_{nm} \end{pmatrix},$$

where r_{ij} stands for the performance evaluation (rating, satisfaction degree) of alternative A_i w.r.t. attribute C_j. The matrix R will be called a decision matrix. Without loss of generality, we assume $r_{ij} \in [0,1]$ for all $i = 1, 2, \ldots, n$ and $j = 1, 2, \ldots, m$ (otherwise, some extra treatment is necessary to make r_{ij} meet the requirement). In addition, the m attributes may have unequal importance. We use w_j to express the importance of C_j, called the weight of C_j, where $w_j \in [0,1]$ $(j = 1, 2, \ldots, m)$ and $\sum_{j=1}^{m} w_j = 1$. Write $W = (w_1, w_2, \ldots, w_m)$. Then a solution of MADM problem is $D = R \circ W^T$, where W^T denotes the transpose of W, or more explicitly,

$$D = \begin{pmatrix} r_{11} & r_{12} & \cdots & r_{1m} \\ r_{21} & r_{22} & \cdots & r_{2m} \\ \cdots & \cdots & & \\ r_{n1} & r_{n2} & \cdots & r_{nm} \end{pmatrix} \circ \begin{pmatrix} w_1 \\ w_2 \\ \vdots \\ w_m \end{pmatrix}.$$

If $D = (d_1, d_2, \ldots, d_n)$, then $d_i = \bigvee_{j=1}^{m} (r_{i,j} \wedge w_j)$ $(1 \leq i \leq n)$ represents the overall performance evaluation (rating, satisfaction degree) of A_i after all C_1, $C_2, \ldots, C_m$ are considered. The final decision is based on the comparison of $d_1, d_2, \ldots, d_n$.

Example 3.14. *To choose a car, we consider the attributes: price(C_1), safety(C_2), gas mileage(C_3) and comfort(C_4). Suppose that there are three types of cars A_1, A_2, and A_3 available and the corresponding decision matrix is:*

$$R = \begin{pmatrix} 0.3 & 0.2 & 1 & 0.2 \\ 0.8 & 1 & 0.4 & 0.3 \\ 0.7 & 0.3 & 0.5 & 0.6 \end{pmatrix}.$$

The number 0.3 in the first row and the first column can be interpreted as the satisfaction degree of A_1 as far as price is concerned. The interpretation is similar for the other numbers. If the attributes C_1, C_2, C_3, C_4 are assigned the weights 0.3, 0.4, 0.1, 0.2 respectively. Then we have

$$D = \begin{pmatrix} 0.3 & 0.2 & 1 & 0.2 \\ 0.8 & 1 & 0.4 & 0.3 \\ 0.7 & 0.3 & 0.5 & 0.6 \end{pmatrix} \circ \begin{pmatrix} 0.3 \\ 0.4 \\ 0.1 \\ 0.2 \end{pmatrix} = \begin{pmatrix} 0.3 \\ 0.4 \\ 0.3 \end{pmatrix}.$$

So A_2 is the best if we take all the attributes into account.

We have to point out that the previous method is just one of many solutions and its effect is context-dependent. Like any other decision making approach, it is not suitable in some cases. According to our experience, the fuzzy method sometimes fills a gap when the traditional method cannot be applied. For instance, there is no traditional method which can be used to represent the fuzziness in a pattern recognition problem as the fuzzy pattern recognition method does. However, in many other cases, it merely provides an alternative method which stands along with other methods to attack practical problems. Fuzzy clustering analysis and the multiple attribute model introduced in this section belong to this category. For more methods and applications with fuzzy multi-objective group decision making, the reader is referred to [96].

3.9 Exercises

1. If $R_1 = \begin{pmatrix} 0.2 & 0.5 & 0.1 \\ 0.8 & 0.2 & 0.9 \\ 0.1 & 0.4 & 0.7 \end{pmatrix}$, $R_2 = \begin{pmatrix} 0.8 & 0.5 & 0.5 \\ 0.4 & 0.3 & 0.7 \\ 1 & 0.5 & 0.8 \end{pmatrix}$,

    what are $R_1 \cup R_2$, $(R_1 \cap R_2)^c$ and $R_1^{-1} \cup R_2$?

2. Let $X = \{x_1, x_2, x_3\}$, $Y = \{y_1, y_2, y_3, y_4\}$. If $R \in F(X \times Y)$ is defined by

    $$R = \begin{pmatrix} 0.8 & 0.6 & 0.7 & 0.4 \\ 0.5 & 0.7 & 1 & 0.5 \\ 0.6 & 0.2 & 0.4 & 0.3 \end{pmatrix},$$

    find $x_i R$ $(i = 1, 2, 3)$, Ry_j $(j = 1, 2, 3, 4)$, $R_{0.5}$ and $R_{0.7}$.

3. Show that for every fuzzy relation R and $\alpha \in [0, 1]$,

    $$(R_\alpha)^{-1} = (R^{-1})_\alpha, \quad (R_{\dot\alpha})^{-1} = (R^{-1})_{\dot\alpha}.$$

4. Let R_1 and R_2 be fuzzy relations on $\mathbb{R}$ defined by

    $$\forall x, y \in \mathbb{R}, R_1(x, y) = e^{-(x-y)^2} \text{ and } R_2(x, y) = e^{y-x}.$$

    Find $(R_1 \cup R_2)^c(3, 2)$ and $(R_1 \cap (R_1^c \cup R_2))(3, 2)$.

5. Let R_1, R_2 be defined as in Exercise 1. Find $R_1 \cup_{S_L} R_2$, $(R_1 \cup_{T_L} R_2)^c$ and $R_1^{-1} \cup_{S_L} R_2$.

6. Let $R \in F(X \times Y)$ and $S \in F(Y \times Z)$ be defined by
$$R = \begin{pmatrix} 0.6 & 0.5 & 0.4 & 0.7 \\ 0.7 & 0.4 & 0.5 & 0.4 \\ 0.5 & 0.6 & 0.4 & 0.8 \end{pmatrix}, \quad S = \begin{pmatrix} 0.7 & 0.4 \\ 0.3 & 0.6 \\ 0.3 & 0.8 \\ 1 & 0.4 \end{pmatrix}.$$
Find $R \circ S$, $R \circ S^c$, $R_{0.6} \circ S_{0.6}$.

7. If R is defined by $R(x,y) = e^{-k(x-y)^2} (k > 0)$, find R^2.

8. Let $R \in F(X \times Y)$ and $S \in F(Y \times Z)$. Show that
$$R \circ S = \bigcup_{\alpha \in [0,1]} \alpha(R_\alpha \circ S_\alpha) = \bigcup_{\alpha \in [0,1]} \alpha(R_{\underline{\alpha}} \circ S_{\underline{\alpha}}).$$

9. If T is a left continuous t-norm, show that:
    (1) $(R_1 \circ_T R_2) \circ_T R_3 = R_1 \circ_T (R_2 \circ_T R_3)$;
    (2) $R \circ_T (\bigcup_{i \in I} R_i) = \bigcup_{i \in I} (R \circ_T R_i)$.

10. Show that $(R_1 \cap R_2) * R_3 = (R_1 * R_3) \cap (R_2 * R_3)$.

11. Show that $(R_1 \circ R_2)^c = R_1^c * R_2^c$ and $(R_1 * R_2)^c = R_1^c \circ R_2^c$.

12. Let (T, S, n) be a De Morgan triple with N a strong negation. Show that if $I(x,y) = S(n(x), y)$, then $R_1 \triangleleft_I (R_2 \triangleleft_I R_3) = (R_1 \circ_T R_2) \triangleleft_I R_3$ and $R_1 \triangleright_I (R_2 \triangleright_I R_3) = R_1 \triangleright_I (R_2 \circ_T R_3)$.

13. Show that if I satisfies $I(x, y) = I(n(y), n(x))$, where n is a negation, then $R_1 \square_E R_2 = (R_1)_n^c \square_E (R_2)_n^c$.

14. If R is a reflexive fuzzy relation on X, prove that $R^2 \supseteq R$ and $\forall x, y \in X$,
$$t(R)(x,y) = \lim_{n \to +\infty} R^n(x,y).$$

15. If R_1 and R_2 are symmetric fuzzy relations, then $R_1 \circ R_2$ is symmetric iff $R_1 \circ R_2 = R_2 \circ R_1$.

16. Let T be a left continuous t-norm. Show that the R-implication I_T is a transitive fuzzy relation on $[0, 1]$.

17. Show that every fuzzy relation is the union of some transitive fuzzy relations.

18. Show that R is transitive iff for any positive integers n, k, $R^{n+k} \subseteq R^n$.

19. Let R be a transitive fuzzy relation on X. Define a crisp relation R' on X by $R' = \{(x,y) | R(x,y) > R(y,x)\}$. Show that R' is transitive.

20. Let the fuzzy relation on $X = [0, +\infty[$ be defined by
$$R(x,y) = \begin{cases} 1 & \text{if } x = y \\ e^{-\max\{x,y\}} & \text{otherwise} \end{cases}.$$

Prove that R is a fuzzy equivalence relation.

21. Let $R_i (i \in I)$ be fuzzy equivalence relations on X. Show that $\bigcap_{i \in I} R_i$ is also a fuzzy equivalence relation on X.

22. Let R_1 and R_2 be T-transitive. Show that
    (1) $R_1 \cap_T R_2$ is T-transitive;
    (2) $R_1 \cap R_2$ is T-transitive.

23. Let R_1, R_2 and R_3 be three fuzzy relations on $X = \{x_1, x_2, x_3, x_4\}$ defined by
$$R_1 = \begin{pmatrix} 1 & 0.3 & 0 & 0.3 \\ 0.4 & 1 & 0.9 & 1 \\ 0 & 0.4 & 1 & 0 \\ 0.7 & 0.9 & 0.7 & 1 \end{pmatrix} \quad R_2 = \begin{pmatrix} 1 & 0.1 & 0.3 & 0.8 \\ 0.1 & 1 & 0.2 & 0.5 \\ 0.3 & 0.2 & 1 & 0.3 \\ 0.8 & 0.5 & 0.3 & 1 \end{pmatrix}$$
$$R_3 = \begin{pmatrix} 1 & 0 & 0 & 0.4 \\ 0 & 1 & 0.9 & 0 \\ 0 & 0.9 & 1 & 0 \\ 0.4 & 0 & 0 & 1 \end{pmatrix}.$$
Are they fuzzy equivalence relations? For every fuzzy equivalence relation, find the partitions of X by R_α ($0 < \alpha \leq 1$).

24. Let $H : [0, 1] \to X \times X$ be such that
    (i) $H(\alpha)$ is an equivalence relation for every $\alpha \in [0, 1]$;
    (ii) $H(\alpha_2) \subseteq H(\alpha_1)$ whenever $\alpha_1 < \alpha_2$. Show that $R = \bigcup_{\alpha \in [0,1]} \alpha H(\alpha)$ is a fuzzy equivalence relation.

25. Show that the reflexive closure of a fuzzy relation R on X is $R \cup I$, where I is the identity relation defined by $I(x, y) = \begin{cases} 1 & x = y \\ 0 & \text{otherwise} \end{cases}$

26. Show that the symmetric closure of a fuzzy relation R on X is $R \cup R^{-1}$.

27. An interior of a fuzzy relation R on X is the greatest relation contained in R. Show that interior is unique if it exists. Denote the interior of R by $\text{Int}(R)$. Show that $\text{Int}(R) = (P(R^c))^c$.

28. If R is a fuzzy relation on a finite universe X, prove that $\forall \alpha \in [0, 1]$, $(t(R))_\alpha = t(R_\alpha)$.

29. If
$$R = \begin{pmatrix} 0.4 & 0.6 \\ 0.7 & 0.5 \end{pmatrix},$$
find $t(R)$. Use $t(R)$ to illustrate whether R is transitive or not.

30. Define the Hamming distance of two fuzzy relations R_1 and R_2 on a finite universe X by:
$$d(R_1, R_2) = \sum_{a,b \in X} |R_1(a, b) - R_2(a, b)|.$$
If R is a fuzzy relation on the finite universe X, show that
$$d(t(R), R) = \min\{d(R', R) | R' \supseteq R, R' \text{ is transitive }\}.$$

3.9 Exercises

31. Let R be a fuzzy relation on a finite universe. Let $R_1 = R$, $R_2 = R_1 \cup R_1^2$, $\ldots$, $R_n = R_{n-1} \cup R_{n-1}^2$, $\ldots$. Show that there exists a positive integer N such that $t(R) = R_N$.

32. If
$$R = \begin{pmatrix} 1 & 0.4 & 0.9 & 0.6 \\ 0.4 & 1 & 0.7 & 0.5 \\ 0.9 & 0.7 & 1 & 0.8 \\ 0.6 & 0.5 & 0.8 & 1 \end{pmatrix},$$
find $t(R)$.

33. Let $R = (r_{ij})_{n \times n}$ be a fuzzy tolerance relation.

    (1) Show that R is a fuzzy equivalence relation when $n = 2$.
    (2) When $n = 3$, suppose that
    $$R = \begin{pmatrix} 1 & \alpha & \beta \\ \alpha & 1 & \gamma \\ \beta & \gamma & 1 \end{pmatrix}.$$
    Show that if R is a fuzzy equivalence relation, then at least two of α, β and γ are identical.
    (3) More generally, verify the following result: If R is a fuzzy equivalence relation on X and $x, y, z \in X$, then at least two of $R(x, y)$, $R(x, z)$ and $R(y, z)$ are identical.

34. Let R be a relation on X. Show that:

    (i) R is complete iff $R \cup R^{-1} \supseteq X \times X \setminus \{(x, x) | x \in X\}$;
    (ii) R is antisymmetric iff $R \cap R^{-1} \subseteq \{(x, x) | x \in X\}$.

35. Let R be a relation on X. Prove the following equivalencies:

    (i) R is negatively transitive iff $\chi_R(a, c) \leq \chi_R(a, b) \vee \chi_R(b, c)$ holds for any $a, b, c \in X$;
    (ii) R is semitransitive iff $\chi_R(a, c) \wedge \chi_R(c, b) \leq \chi_R(a, d) \vee \chi_R(d, b)$ holds for any $a, b, c, d \in X$;
    (iii) R is a Ferrers relation iff $\chi_R(a, b) \wedge \chi_R(c, d) \leq \chi_R(a, d) \vee \chi_R(c, b)$ holds for any $a, b, c, d \in X$.

36. Show that:

    (i) R is negatively S-transitive iff $R \subseteq R *_S R$;
    (ii) R is T-S semitransitive iff $R \circ_T R \subseteq R *_S R$.

37. In Proposition 3.22, if condition ($\mathcal{C}$) is replaced by the condition that T has no zero divisor, i.e. $\forall x, y \in]0, 1], T(x, y) > 0$, show that the involved conclusions are still valid.

38. Give an example to illustrate that the condition ($\mathcal{C}$) is not generally true for T_0 and S_0.

39. Show that both a strong De Morgan triple and a strict De Morgan triple meet the condition ($\mathcal{C}$).
40. Show that a fuzzy relation R on X is consistent iff $(\forall a, b, c \in X)$, $[(R(a,b) \geq R(b,a)$ and $R(b,c) \geq R(c,b)$ with strict inequality holding at least once) imply $R(a,c) > R(c,a)$.]
41. If $(\forall a, b, c \in X)$, $(R(a,b) \geq R(b,a)$ implies $R(a,c) \geq R(b,c)$ and $R(c,b) \geq R(c,a))$, then R is called strongly transitive. Show that strong transitivity implies consistency.
42. Show that if a fuzzy relation R on X is transitive and not consistent, then there exist $a, b, c \in X$ such that $R(a,b) = R(b,a) = R(b,c) = R(c,b)$ and $R(a,c) < R(c,a)$.
43. Show that R is consistent iff R is acyclic and $(\forall a, b, c \in X)$, $(R(a,b) > R(b,a)$ and $R(b,c) = R(c,b)$ imply $R(a,c) > R(c,a))$.
44. Consider the statements:

    (1) $(\forall a, b, c \in X)$, $(R(a,b) = R(b,a)$ and $R(b,c) = R(c,b)$ imply $R(a,c) = R(c,a))$;
    (2) $(\forall a, b, c \in X)$, $(R(a,b) > R(b,a)$ and $R(b,c) = R(c,b)$ imply $R(a,c) > R(c,a))$.

    We have the following conclusions:

    (i) If R is weakly transitive, then (1) implies (2);
    (ii) (2) implies (1).

45. Determine the scope of λ, λ_1 and λ_2 in the following equations provided that they are consistent.

$$\begin{pmatrix} 0.4 & 0.2 & 0.2 \\ 0.2 & \lambda & 0.1 \\ 0.1 & 0.3 & 0.3 \\ 0.8 & 0.7 & 0.8 \end{pmatrix} \circ X = \begin{pmatrix} 0.3 \\ 0.4 \\ 0.3 \\ 0.7 \end{pmatrix}$$

$$\begin{pmatrix} \lambda_1 & 0.7 & 0.3 & 0.3 \\ 0.2 & 0.3 & 0.2 & \lambda_2 \\ 0.1 & 0.5 & 0.4 & 0.5 \\ 0.8 & 0.2 & 0 & 0.4 \end{pmatrix} \circ X = \begin{pmatrix} 0.4 \\ 0.2 \\ 0.5 \\ 0.4 \end{pmatrix}$$

46. Solve the following fuzzy relation equations for X:

(1) $\begin{pmatrix} 0.6 & 0 & 0.6 & 0.8 \\ 0.5 & 0.3 & 0 & 0.4 \\ 0.3 & 0.1 & 0.8 & 0.3 \end{pmatrix} \circ X = \begin{pmatrix} 0.2 \\ 0.3 \\ 0.3 \end{pmatrix}$

(2) $\begin{pmatrix} 0.2 & 0.2 & 0 & 0.2 \\ 0.3 & 0.7 & 0.4 & 0.3 \\ 0.2 & 0.1 & 0.5 & 0.4 \\ 0.1 & 0 & 0.1 & 0.1 \end{pmatrix} \circ X = \begin{pmatrix} 0.2 \\ 0.4 \\ 0.3 \\ 0.1 \end{pmatrix}$

3.9 Exercises

(3) $X \circ \begin{pmatrix} 0.3 & 0.6 & 0.1 & 0 \\ 0 & 0.2 & 0.5 & 0.3 \\ 0.5 & 0.3 & 0.1 & 0.1 \\ 0.1 & 0.3 & 0.2 & 0.4 \end{pmatrix} = (0.2, 0.2, 0.4, 0.3)$

(4) $\begin{pmatrix} 0.7 & 0.3 & 0.7 \\ 0.2 & 0.5 & 0.6 \end{pmatrix} \circ X = \begin{pmatrix} 0.7 & 0.4 \\ 0.5 & 0.5 \end{pmatrix}$

(5) $X \circ \begin{pmatrix} 0.7 & 0.5 & 0.4 \\ 0.5 & 0.6 & 0.3 \\ 0.6 & 0.7 & 0.4 \end{pmatrix} = \begin{pmatrix} 0.5 & 0.6 & 0.4 \\ 0.5 & 0.5 & 0.3 \\ 0.6 & 0.7 & 0.4 \end{pmatrix}$

47. For any real numbers a, b, write $a(b) = \begin{cases} 1 & a > b \\ b & a = b \\ 0 & a < b \end{cases}$.

    Let $R = (r_{ij})_{n \times m}$, $B = (b_1, b_2, \ldots, b_m)$, and $R^* = (r_{ij}(b_j))_{n \times m}$. Prove that the equations $X \circ R = B$ and $X \circ R^* = B$ have the same set of solutions.

48. Let T be a t-norm and $R \in V \times W$ and $S \in U \times W$. Consider the fuzzy relation equation $X \circ_T R = S$. Define $\overline{X} \in U \times V$ by $\forall (u, v) \in U \times V$, $\overline{X}(u, v) = \inf_w I_T(R(v, w), S(u, w))$. Show that if T is left continuous, then $\overline{X}$ is the maximal solution of $X \circ_T R = S$.

49. Given $X = \{x_1, x_2, \cdots, x_7\}$, use the direct clustering method to obtain the equivalency classes of the following fuzzy relation at different levels.

$$R = \begin{pmatrix} 1 & 0.9 & 0.7 & 0.5 & 0.9 & 1 & 0.4 \\ 0.9 & 1 & 0.5 & 0.4 & 0.5 & 0.4 & 0.1 \\ 0.7 & 0.5 & 1 & 0.5 & 0.7 & 0.1 & 0.4 \\ 0.5 & 0.4 & 0.5 & 1 & 0.5 & 0.7 & 0.9 \\ 0.9 & 0.5 & 0.7 & 0.5 & 1 & 0.7 & 0.1 \\ 1 & 0.4 & 0.1 & 0.7 & 0.7 & 1 & 0.5 \\ 0.4 & 0.1 & 0.4 & 0.9 & 0.1 & 0.5 & 1 \end{pmatrix}.$$

50. Given $X = \{x_1, x_2, \cdots, x_7\}$, use the direct clustering method to obtain the equivalency class of the following fuzzy relation at the level 0.52.

$$R = \begin{pmatrix} 1 & 0.1 & 0.6 & 0.3 & 0.4 & 0.5 & 0.8 \\ 0.1 & 1 & 0.1 & 0.8 & 0.7 & 0.5 & 0.1 \\ 0.6 & 0.1 & 1 & 0.2 & 0.5 & 0.3 & 0.8 \\ 0.3 & 0.8 & 0.2 & 1 & 0.4 & 0.6 & 0.2 \\ 0.4 & 0.7 & 0.5 & 0.4 & 1 & 0.4 & 0.4 \\ 0.5 & 0.5 & 0.3 & 0.6 & 0.4 & 1 & 0.3 \\ 0.8 & 0.1 & 0.8 & 0.2 & 0.4 & 0.3 & 1 \end{pmatrix}.$$

Chapter 4
Extension Principle and Fuzzy Numbers

In the traditional multi-attribute decision analysis, there is a well-defined problem-solving model–the Simple Additive Weighting (SAW) method. This model can be formulated as follows. Let $A_1, A_2, \ldots, A_n$ be n alternatives and $C_1, C_2, \ldots, C_m$ m attributes with the corresponding weights $w_1, w_2, \ldots, w_m$ respectively. If the evaluation of the alternative A_i w.r.t. C_j is denoted by r_{ij}, then the overall evaluation of A_i may be computed as

$$r_i = \frac{\sum_{j=1}^{m} w_j r_{ij}}{\sum_{j=1}^{m} w_j} \quad (i = 1, 2, \cdots, n).$$

and the final ranking of alternatives $A_1, A_2, \ldots, A_n$ is based on the comparison of the real numbers $r_1, r_2, \ldots, r_n$. In a real world problem, it is often difficult to give an evaluation for the weights or the attributes in a precise way. Sometimes it is more reasonable to say that 'the evaluation is approximately 0.7' than 'the evaluation is exactly 0.7'. Such situation arises typically due to the lack of objective information or/and the use of subjective estimations. The traditional mathematical theory can give little help as to make a decision with such kind of imperfect knowledge. On the contrast, fuzzy set theory provides a strongly effective apparatus to deal with the problem. Firstly, imprecise descriptions can be modeled by means of fuzzy numbers. That is to say, the crisp numbers r_{ij} or w_j $(i = 1, 2, \ldots, n; j = 1, 2, \ldots, m)$ or both of them may be replaced by fuzzy numbers $\tilde{r}_{ij}$ and $\tilde{w}_j$ respectively. Accordingly, the overall evaluation of A_i becomes

$$\tilde{r}_i = \frac{\sum_{j=1}^{m} \tilde{w}_j \tilde{r}_{ij}}{\sum_{j=1}^{m} \tilde{w}_j} \quad (i = 1, 2, \cdots, n).$$

Furthermore, operations like the weighted sum of fuzzy numbers can also be formulated in the framework of fuzzy set theory using the Zadeh's extension principle. In this way, we can still evaluate the performance of alternatives although only imprecise data are available. This is significant particularly when modeling a system for which there is no way or it is simply not necessary to acquire the full knowledge of it. After $\tilde{r}_1, \tilde{r}_2, \cdots, \tilde{r}_n$, which are still fuzzy numbers, are obtained, the problem of how to rank them arises. Unlike in the crisp case, fuzzy numbers have no natural order. As a result, we have to design a way to compare the fuzzy numbers $\tilde{r}_1, \tilde{r}_2, \cdots, \tilde{r}_n$ in order to rank the alternatives $A_1, A_2, \ldots, A_n$. It is the so-called problem of ranking fuzzy numbers. The same problem is encountered in Fuzzy Decision Tree [2, 158], Fuzzy Analytic Hierarchy Process [19, 87, 145], Fuzzy Linear Programming [36, 125], even electrocardiological diagnostics [75] after the fuzzy information is involved and processed. It is because of its extensive applications that many efforts have been made to deal with the problem.

Therefore, three tasks will be fulfilled in this chapter. The first is the introduction of the Zadeh's extension principle. The second is the discussion of fuzzy numbers and their algebraic operations. The last one is the detailed investigation of ranking methods for fuzzy numbers.

4.1 Unary Extension Principle

Let f be a mapping from X to Y. We recall the (direct) image of a subset A of X under f is defined as

$$f(A) = \{f(x) | x \in A\} \subseteq Y$$

and the inverse image of a subset B of Y is defined as

$$f^{-1}(B) = \{x | f(x) \in B\} \subseteq X.$$

In other words, the mapping f from X to Y induces two new mappings $f : P(X) \to P(Y)$ and $f^{-1} : P(Y) \to P(X)$. These induced mappings possess the following properties.

Proposition 4.1. *If $f : X \to Y$, $A, B, A_i \in P(X)$ ($i \in I$), then the induced mapping f satisfies that:*

(1) $A \subseteq B$ implies $f(A) \subseteq f(B)$;
(2) $f(\bigcup_{i \in I} A_i) = \bigcup_{i \in I} f(A_i)$;
(3) $f(\bigcap_{i \in I} A_i) \subseteq \bigcap_{i \in I} f(A_i)$;
(4) $\forall y \in Y$, $(f(A))(y) = \bigvee_{f(x)=y} A(x)$.

4.1 Unary Extension Principle

Proof. (1) Trivial.

(2) Let $y \in Y$. Then

$$y \in f(\bigcup_{i \in I} A_i) \Leftrightarrow \exists x \in \bigcup_{i \in I} A_i, f(x) = y$$
$$\Leftrightarrow \exists i \in I, x \in A_i \text{ and } y = f(x)$$
$$\Leftrightarrow \exists i \in I, y \in f(A_i)$$
$$\Leftrightarrow y \in \bigcup_{i \in I} f(A_i).$$

(3) Similar to the proof of (2).

(4) Let $y \in Y$. Then

$$(f(A))(y) = 1 \Leftrightarrow y \in f(A)$$
$$\Leftrightarrow \exists x \in A, y = f(x)$$
$$\Leftrightarrow \exists x \in X, A(x) = 1 \text{ and } y = f(x)$$
$$\Leftrightarrow \bigvee_{f(x)=y} A(x) = 1. \qquad \square$$

Proposition 4.2. *If $f : X \to Y$, $A, B, B_j \in P(Y)$ ($j \in J$), then the induced mapping f^{-1} satisfies that:*

(1) $A \subseteq B$ implies $f^{-1}(A) \subseteq f^{-1}(B)$;
(2) $f^{-1}(\bigcup_{j \in J} B_j) = \bigcup_{j \in J} f^{-1}(B_j)$;
(3) $f^{-1}(\bigcap_{j \in J} B_j) = \bigcap_{j \in J} f^{-1}(B_j)$;
(4) $\forall x \in X, (f^{-1}(B))(x) = B(f(x))$.

Proof. The proof is left as an exercise (Exercise 1). $\qquad \square$

Definition 4.1. *(Zadeh's extension principle) Let f be a mapping from X to Y. Then f induces two new mappings $f : F(X) \to F(Y)$, defined by*

$$\forall y \in Y, \forall A \in F(X), (f(A))(y) = \begin{cases} \bigvee_{f(x)=y} A(x), & y \in f(X) \\ 0 & \text{otherwise} \end{cases}$$

and $f^{-1} : F(Y) \to F(X)$, defined by

$$\forall x \in X, \forall B \in F(Y), (f^{-1}(B))(x) = B(f(x)).$$

$f(A)$ is understood as the (direct) image of the fuzzy set A on X under f and $f^{-1}(B)$ is the inverse image of the fuzzy set B on Y under f. Hence, the

Zadeh's extension principle extends the notions of image and inverse image to the fuzzy case.

Example 4.1. Let $X = \{x_1, x_2, x_3, x_4, x_5, x_6\}$ and $Y = \{a, b, c, d\}$. A mapping f from X to Y and a fuzzy set A on X are defined by:

$$f(x) = \begin{cases} a & x \in \{x_1, x_2, x_3\}; \\ b & x \in \{x_4, x_5\}; \\ c & x = x_6. \end{cases}$$

$$A = 1/x_1 + 0.5/x_2 + 0.8/x_4 + 0.4/x_5 + 0.7/x_6.$$

Applying the extension principle yields

$$(f(A))(a) = \bigvee_{f(x)=a} A(x) = A(x_1) \vee A(x_2) \vee A(x_3) = 1.$$

Similarly,

$$(f(A))(b) = 0.8, (f(A))(c) = 0.7 \text{ and } (f(A))(d) = 0.$$

Hence

$$f(A) = 1/a + 0.8/b + 0.7/c.$$

In addition, let $B = f(A)$. Then

$$f^{-1}(B)(x_1) = B(f(x_1)) = f(A)(a) = 1.$$

Similarly,

$$f^{-1}(B)(x_2) = f^{-1}(B)(x_3) = 1, \; f^{-1}(B)(x_4) = f^{-1}(B)(x_5) = 0.8, \text{ and}$$

$f^{-1}(B)(x_6) = 0.7$. *Hence,*

$$f^{-1}(f(A)) = 1/x_1 + 1/x_2 + 1/x_3 + 0.8/x_4 + 0.8/x_5 + 0.7/x_6.$$

By Example 4.1, $f^{-1}(f(A)) = A$ is not necessarily valid. Instead, it can be generally verified that:

(1) $\forall A \in F(X), f^{-1}(f(A)) \supseteq A$;
(2) $\forall B \in F(Y), f(f^{-1}(B)) \subseteq B$.

For more specific results for injective and surjective mappings, see Exercise 3.

Example 4.2. Let $f(x) = x^2$ and $A =$ *around* 1 be defined by

$$A(x) = \begin{cases} x & 0 \leq x \leq 1 \\ 2 - x & 1 < x \leq 2 \\ 0 & \text{otherwise} \end{cases}$$

4.1 Unary Extension Principle

Then $\forall y \in [0, +\infty[$,

$$f(A)(y) = \bigvee_{f(x)=y} A(x) = \bigvee_{x^2=y} A(x).$$

$$f(A)(y) = \bigvee_{x^2=y} A(x) = A(\sqrt{y}) \vee A(-\sqrt{y})$$

$$= A(\sqrt{y}) = \begin{cases} \sqrt{y} & y \in [0,1] \\ 2 - \sqrt{y} & y \in]1,4] \\ 0 & y > 4 \end{cases}.$$

When $y \in]-\infty, 0[$, $f(A)(y) = 0$.

Since $f(x) = x^2$, we can define A^2 as $f(A)$. In this sense, the square of around 1 is found. More generally, all unary operations on $\mathbb{R}$ may be extended using the extension principle.

Proposition 4.3. *If $f : X \to Y$, $A, B, A_i \in F(X)$ $(i \in I)$, then*

(1) $A \subseteq B$ implies $f(A) \subseteq f(B)$;
(2) $f(\bigcup_{i \in I} A_i) = \bigcup_{i \in I} f(A_i)$;
(3) $f(\bigcap_{i \in I} A_i) \subseteq \bigcap_{i \in I} f(A_i)$;
(4) $\forall \alpha \in]0, 1]$, $f(A_\alpha) \subseteq (f(A))_\alpha$;
(5) $\forall \alpha \in [0, 1[$, $(f(A))_{\underline{\alpha}} = f(A_{\underline{\alpha}})$;
(6) $f(A) = \bigcup_{\alpha \in [0,1]} \alpha f(A_\alpha) = \bigcup_{\alpha \in [0,1]} \alpha f(A_{\underline{\alpha}})$.

Proof. We prove (2), (5) and (6) as examples.
(2) For every $y \in Y$,

$$(f(\bigcup_{i \in I} A_i))(y) = \bigvee_{f(x)=y} (\bigcup_{i \in I} A_i)(x)$$

$$= \bigvee_{f(x)=y} \bigvee_{i \in I} A_i(x) = \bigvee_{i \in I} \bigvee_{f(x)=y} A_i(x)$$

$$= \bigvee_{i \in I} (f(A_i))(y) = (\bigcup_{i \in I} f(A_i))(y).$$

(5) Let $y \in Y$. Then

$$y \in (f(A))_{\underline{\alpha}} \Leftrightarrow (f(A))(y) > \alpha$$

$$\Leftrightarrow \bigvee_{f(x)=y} A(x) > \alpha$$

$$\Leftrightarrow \exists x \in X, f(x) = y, A(x) > \alpha$$

$$\Leftrightarrow \exists x \in A_{\underline{\alpha}}, f(x) = y$$

$$\Leftrightarrow y \in f(A_{\underline{\alpha}}).$$

(6) By Decomposition Theorem II and (5),

$$f(A) = \bigcup_{\alpha \in [0,1]} \alpha(f(A))_{\underline{\alpha}} = \bigcup_{\alpha \in [0,1]} \alpha f(A_{\underline{\alpha}}).$$

For every y in Y,

$$\left(\bigcup_{\alpha \in [0,1]} \alpha f(A_\alpha)\right)(y) = \bigvee_{\alpha \in [0,1]} (\alpha \wedge (f(A_\alpha))(y))$$

$$= \bigvee_{\alpha \in [0,1]} \left(\alpha \wedge \bigvee_{f(x)=y} A_\alpha(x)\right)$$

$$= \bigvee_{\alpha \in [0,1]} \bigvee_{f(x)=y} (\alpha \wedge A_\alpha(x))$$

$$= \bigvee_{f(x)=y} \bigvee_{\alpha \in [0,1]} (\alpha \wedge A_\alpha(x))$$

$$= \bigvee_{f(x)=y} A(x) = (f(A))(y).$$

Thus $f(A) = \bigcup_{\alpha \in [0,1]} \alpha f(A_\alpha)$. □

It should be noted that $f(A_\alpha) = (f(A))_\alpha$ does not hold generally.

Example 4.3. *Assume that $X = \{x_1, x_2, \cdots, x_n, \cdots\}$ and $Y = \{0\}$. The fuzzy set A on X and the mapping $f : X \to Y$ are respectively defined by $A(x_n) = 1 - \frac{1}{n}$ and $f(x_n) = 0$ $(n = 1, 2, \cdots)$. In this case, $f(A_1) = \emptyset$ and $[f(A)]_1 = Y$, and thus $f(A_1) \neq [f(A)]_1$.*

Now the question naturally arises: when the equality holds for all $\alpha \in]0, 1]$? To answer this question, we need the following notion. Given $a_i \in [0, 1]$ $(i \in I)$, we say that $\sup\{a_i | i \in I\}$ is attained if there exists an $i_0 \in I$ such that $\sup\{a_i | i \in I\} = a_{i_0}$.

Proposition 4.4. *Let A be a fuzzy set on X. Then $\forall \alpha \in]0, 1]$, $(f(A))_\alpha = f(A_\alpha) \Leftrightarrow (f(A))(y)$ is attained for every y in $f(X)$.*

Proof. "$\Leftarrow$" part. Let $\alpha \in]0, 1]$. If $y \in (f(A))_\alpha$, then $(f(A))(y) \geq \alpha$. Noticing that $\alpha > 0$, we have $y \in f(X)$ (Otherwise, there is no $x \in X$ such that $f(x) = y$, and thus $(f(A))(y) = 0 < \alpha$, a contradiction). Since $(f(A))(y)$ is attained, there exists an x such that $f(x) = y$ and $(f(A))(y) = A(x)$. Hence $x \in A_\alpha$, whence $y \in f(A_\alpha)$. As a consequence, $(f(A))_\alpha \subseteq f(A_\alpha)$. Now $(f(A))_\alpha = f(A_\alpha)$ follows from Proposition 4.3(4).

"$\Rightarrow$" part. For every y in $f(X)$, let $\alpha = (f(A))(y)$. Then $y \in (f(A))_\alpha = f(A_\alpha)$. There exists an x in A_α such that $y = f(x)$, and thus $A(x) \geq \alpha = (f(A))(y)$. On the other hand, $(f(A))(y) \geq A(x)$ since $y = f(x)$. Therefore $(f(A))(y) = A(x)$, i.e. $(f(A))(y)$ is attained for every y in $f(X)$. □

Properties of the inverse image of fuzzy sets are much better as can be seen in the following proposition.

Proposition 4.5. *Let $f : X \to Y$ and $A, B, B_i \in F(Y)$ ($i \in I$). Then the induced mapping f^{-1} satisfies that:*

(1) $A \subseteq B$ implies that $f^{-1}(A) \subseteq f^{-1}(B)$;
(2) $f^{-1}(B^c) = (f^{-1}(B))^c$;
(3) $f^{-1}(\bigcup_{i \in I} B_i) = \bigcup_{i \in I} f^{-1}(B_i)$;
(4) $f^{-1}(\bigcap_{i \in I} B_i) = \bigcap_{i \in I} f^{-1}(B_i)$;
(5) $\forall \alpha \in]0, 1]$, $f^{-1}(B_\alpha) = (f^{-1}(B))_\alpha$;
(6) $\forall \alpha \in [0, 1[$, $(f^{-1}(B))_{\underset{\sim}{\alpha}} = f^{-1}(B_{\underset{\sim}{\alpha}})$;
(7) $f^{-1}(B) = \bigcup_{\alpha \in [0,1]} \alpha f^{-1}(B_\alpha) = \bigcup_{\alpha \in [0,1]} \alpha f^{-1}(B_{\underset{\sim}{\alpha}})$.

Proof. We prove (4) and (5) as examples.
(4) $\forall x \in X$,

$$f^{-1}(\bigcap_{i \in I} B_i)(x) = (\bigcap_{i \in I} B_i)(f(x)) = \bigwedge_{i \in I} B_i(f(x))$$
$$= \bigwedge_{i \in I} f^{-1}(B_i)(x) = (\bigcap_{i \in I} f^{-1}(B_i))(x).$$

(5) $x \in f^{-1}(B_\alpha) \Leftrightarrow f(x) \in B_\alpha \Leftrightarrow B(f(x)) \geq \alpha \Leftrightarrow (f^{-1}(B))(x) \geq \alpha \Leftrightarrow x \in (f^{-1}(B))_\alpha$. □

4.2 n-Ary Extension Principle

As we know, the cartesian product is a bridge between a single object and multiple objects. The cartesian product of fuzzy sets is a useful tool for extending the unary extension principle to the n-ary case.

Definition 4.2. *Let $A_1, A_2, \cdots, A_n$ be fuzzy sets on $X_1, X_2, \cdots, X_n$ respectively. Define $A_1 \times A_2 \times \cdots \times A_n \in F(X_1 \times X_2 \times \cdots \times X_n)$ by*

$$\forall (x_1, x_2, \cdots, x_n) \in X_1 \times X_2 \times \cdots \times X_n,$$

$$(A_1 \times A_2 \times \cdots \times A_n)(x_1, x_2, \cdots, x_n) = \min(A_1(x_1), A_2(x_2), \cdots, A_n(x_n)).$$

Then $A_1 \times A_2 \times \cdots \times A_n$ is said to be the cartesian product of $A_1, A_2, \cdots, A_n$ and it will be denoted by $\prod_{i=1}^{n} A_i$ for simplicity.

Lemma 4.1. $\forall \alpha \in [0,1]$, $(\prod_{i=1}^{n} A_i)_\alpha = \prod_{i=1}^{n}(A_i)_\alpha$ and $(\prod_{i=1}^{n} A_i)_{\dot\alpha} = \prod_{i=1}^{n}(A_i)_{\dot\alpha}$

Proof. $(x_1, x_2, \cdots, x_n) \in (\prod_{i=1}^{n} A_i)_\alpha \Leftrightarrow (\prod_{i=1}^{n} A_i)(x_1, x_2, \cdots, x_n) \geq \alpha \Leftrightarrow \min(A_1(x_1), A_2(x_2), \cdots, A_n(x_n)) \geq \alpha \Leftrightarrow A_i(x_i) \geq \alpha$ $(i = 1, 2, \cdots, n)$ $\Leftrightarrow x_i \in (A_i)_\alpha$ $(i = 1, 2, \cdots, n) \Leftrightarrow (x_1, x_2, \cdots, x_n) \in \prod_{i=1}^{n}(A_i)_\alpha$. Hence $(\prod_{i=1}^{n} A_i)_\alpha = \prod_{i=1}^{n}(A_i)_\alpha$.

The proof of the other identity is similar. □

Theorem 4.1. $\prod_{i=1}^{n} A_i = \bigcup_{\alpha \in [0,1]} \alpha \prod_{i=1}^{n}(A_i)_\alpha = \bigcup_{\alpha \in [0,1]} \alpha \prod_{i=1}^{n}(A_i)_{\dot\alpha}$.

Proof. By Decomposition Theorem I and Lemma 4.1,

$$\prod_{i=1}^{n} A_i = \bigcup_{\alpha \in [0,1]} \alpha (\prod_{i=1}^{n} A_i)_\alpha = \bigcup_{\alpha \in [0,1]} \alpha \prod_{i=1}^{n}(A_i)_\alpha.$$

Similarly, it follows from Decomposition Theorem II and Lemma 4.1 that

$$\prod_{i=1}^{n} A_i = \bigcup_{\alpha \in [0,1]} \alpha \prod_{i=1}^{n}(A_i)_{\dot\alpha}. \qquad \square$$

Before we proceed to introduce the n-ary fuzzy extension of a mapping on a cartesian product, we define the image of a subset of the cartesian product under a mapping. Let f be a mapping from $X_1 \times X_2 \times \cdots \times X_n$ to Y and $A_i \subseteq X_i$ $(i = 1, 2, \cdots, n)$. By the image of $\prod_{i=1}^{n} A_i$ under f, we mean

$$f(\prod_{i=1}^{n} A_i) = \{f(x_1, x_2, \cdots, x_n) | x_i \in A_i \text{ for } i = 1, 2, \cdots, n\},$$

denoted $f(A_1, A_2, \cdots, A_n)$.

Example 4.4. *The notion of the image of the cartesian product can be employed to define algebraic operations of sets on the real line. Let us consider the addition $[a,b] + [c,d]$ of two closed intervals $[a,b]$ and $[c,d]$ as an example. For this purpose, let $f(x_1, x_2) = x_1 + x_2$. Then define*

$$[a,b] + [c,d] \triangleq f([a,b], [c,d]) = \{f(x_1, x_2) | x_1 \in [a,b], x_2 \in [c,d]\}$$
$$= \{x_1 + x_2 | x_1 \in [a,b], x_2 \in [c,d]\} = [a+c, b+d].$$

Similarly,
$[a,b] - [c,d] = [a-d, b-c]$,
$[a,b] \times [c,d] = [\min(ac, ad, bc, bd), \max(ac, ad, bc, bd)]$,

4.2 n-Ary Extension Principle

$[a,b] \div [c,d] = [\min(a/c, a/d, b/c, b/d), \max(a/c, a/d, b/c, b/d)]$ $(0 \notin [c,d])$.
$[a,b] \vee [c,d] = [a \vee c, b \vee d]$,
$[a,b] \wedge [c,d] = [a \wedge c, b \wedge d]$,

For instance, $[2,3] \div ([-1,0] - 2) = [-3/2, -2/3]$.

Now we state the n-ary extension principle of fuzzy sets. Let $f \colon X_1 \times X_2 \times \cdots \times X_n \to Y$. Then f induces a new mapping, also denoted f, from $F(X_1) \times F(X_2) \times \cdots \times F(X_n)$ to $F(Y)$, defined by

$$\forall A_i \in F(X_i)(i = 1, 2, \cdots, n), f(A_1, A_2, \cdots, A_n) = f(A_1 \times A_2 \times \cdots \times A_n).$$

Theorem 4.2

$$\forall y \in f(X_1, X_2, \cdots, X_n), \forall A_i \in F(X_i)(i = 1, 2, \cdots, n),$$

$$(f(A_1, A_2, \cdots, A_n))(y) = \bigvee_{f(x_1, x_2, \cdots, x_n) = y} \bigwedge_{i=1}^{n} A_i(x_i).$$

Proof. Let $\forall y \in Y$. By definition,

$$f(A_1, A_2, \cdots, A_n)(y) = f(A_1 \times A_2 \times \cdots \times A_n)(y)$$

$$= \bigvee_{f(x_1, x_2, \cdots, x_n) = y} (A_1 \times A_2 \times \cdots \times A_n)(x_1, x_2, \cdots, x_n)$$

$$= \bigvee_{f(x_1, x_2, \cdots, x_n) = y} \bigwedge_{i=1}^{n} A_i(x_i). \qquad \square$$

Example 4.5. *Let two fuzzy sets A_1 and A_2 on $X = \{0, 1, 2, \cdots \cdots\}$ be defined by*

$$A_1 = 0.3/0 + 1/1 + 0.3/2$$
$$A_2 = 0.2/1 + 1/2 + 0.2/3.$$

If $f(x_1, x_2) = x_1 + x_2$, then $A_1 + A_2 \triangleq f(A_1, A_2)$ is obtained by the extension principle as

$$(A_1 + A_2)(y) = \bigvee_{f(x_1, x_2) = y} (A_1(x_1) \wedge A_2(x_2)) = \bigvee_{x_1 + x_2 = y} (A_1(x_1) \wedge A_2(x_2)).$$

Hence

$$(A_1 + A_2)(1) = \bigvee_{x_1 + x_2 = 1} (A_1(x_1) \wedge A_2(x_2))$$
$$= A_1(0) \wedge A_2(1) = 0.2.$$

$$(A_1 + A_2)(2) = \bigvee_{x_1+x_2=2} (A_1(x_1) \wedge A_2(x_2))$$
$$= (A_1(0) \wedge A_2(2)) \vee (A_1(1) \wedge A_2(1)) = 0.3.$$

Similarly,
$$(A_1 + A_2)(3) = 1, (A_1 + A_2)(4) = 0.3, (A_1 + A_2)(5) = 0.2,$$

and
$$(A_1 + A_2)(y) = 0 \text{ for } y \notin \{1, 2, 3, 4, 5\}.$$

Therefore,
$$A_1 + A_2 = 0.2/1 + 0.3/2 + 1/3 + 0.3/4 + 0.2/5.$$

Properties of $f(A_1, A_2, \cdots, A_n)$ are similar to those of $f(A)$ and omitted here. Some of them will appear as exercises (Exercises 11 and 12).

4.3 Convex Fuzzy Quantities

Let us start with the definition of a fuzzy quantity.

Definition 4.3. *A fuzzy set A on the real line $\mathbb{R}$, or $A \in F(\mathbb{R})$, is called a fuzzy quantity. A fuzzy quantity A is called non-negative, denoted $A \geq 0$, if $supp(A) \subseteq [0, +\infty[$. A fuzzy quantity A is called positive, denoted $A > 0$, if $supp(A) \subseteq]0, +\infty[$.*

By definition, every subset of $\mathbb{R}$ is a fuzzy quantity, and thus every real number a, which can be identified with the singleton $\{a\}$ can be also regarded as a fuzzy quantity. In this sense, a fuzzy quantity is an extension of a real number. Arithmetical operations of fuzzy quantities can be performed by applying the Zadeh's extension principle. Through simple calculations as in Example 4.5, we obtain the following formulas for the basic arithmetic operations of fuzzy quantities. Let A_1 and A_2 be fuzzy quantities and $y \in \mathbb{R}$.

$$(A_1 + A_2)(y) = \bigvee_{x \in \mathbb{R}} (A_1(x) \wedge A_2(y-x)).$$

$$(A_1 - A_2)(y) = \bigvee_{x \in \mathbb{R}} (A_1(x) \wedge A_2(x-y)).$$

$$(A_1 A_2)(y) = \bigvee_{xz=y} (A_1(x) \wedge A_2(z)).$$

$$(A_1/A_2)(y) = \bigvee_{x/z=y} (A_1(x) \wedge A_2(z))(0 \notin supp(A_2).$$

4.3 Convex Fuzzy Quantities

$$(A_1 \vee A_2)(y) = \bigvee_{x \vee z = y} (A_1(x) \wedge A_2(z)).$$

$$(A_1 \wedge A_2)(y) = \bigvee_{x \wedge z = y} (A_1(x) \wedge A_2(z)).$$

The concept of a fuzzy quantity is very general. As a result, arithmetical operations of fuzzy quantities fulfil merely some simple properties, e.g. $A_1 + A_2 = A_2 + A_1$, $A_1 A_2 = A_2 A_1$, $A_1 \vee A_2 = A_2 \vee A_1$, $A_1 \wedge A_2 = A_2 \wedge A_1$, $A_1 + (A_2 + A_3) = (A_1 + A_2) + A_3$, $A_1(A_2 A_3) = (A_1 A_2) A_3$, etc.. Most properties of real numbers are not satisfied any more for fuzzy quantities. To improve their performance, we shall impose some restrictions on them. The first one is convexity.

Recall that a subset A on a linear space X is convex iff $\forall x_1, x_2 \in A$ implies $\forall \alpha \in [0,1]$, $\alpha x_1 + (1-\alpha) x_2 \in A$. To figure out the definition of a convex fuzzy set, consider the following description of the characteristic function of a convex set.

$$\begin{aligned}
A \text{ is convex} &\Leftrightarrow (\forall x_1, x_2 \in A \Rightarrow \forall \alpha \in [0,1], \alpha x_1 + (1-\alpha) x_2 \in A) \\
&\Leftrightarrow (\forall x_1, x_2 \in X, A(x_1) = 1 \text{ and } A(x_2) = 1 \\
&\quad \Rightarrow \forall \alpha \in [0,1], A(\alpha x_1 + (1-\alpha) x_2) = 1) \\
&\Leftrightarrow (\forall x_1, x_2 \in X, A(x_1) \wedge A(x_2) = 1 \\
&\quad \Rightarrow \forall \alpha \in [0,1], A(\alpha x_1 + (1-\alpha) x_2) = 1) \\
&\Leftrightarrow (\forall x_1, x_2 \in X, \forall \alpha \in [0,1], \\
&\quad A(x_1) \wedge A(x_2) \leq A(\alpha x_1 + (1-\alpha) x_2)).
\end{aligned}$$

Now the following definition becomes natural.

Definition 4.4. *Let A be a fuzzy set on a linear space X. A is called a convex fuzzy set on X if, whenever $x_1, x_2 \in X$ and $\alpha \in [0,1]$,*

$$A(\alpha x_1 + (1-\alpha) x_2) \geq \min(A(x_1), A(x_2)).$$

Proposition 4.6. *A is a convex fuzzy set on X iff for every $\lambda \in [0,1]$, A_λ is a convex set on X.*

Proof. Necessity. Assume that A is a convex fuzzy set. If $x_1, x_2 \in A_\lambda$ and $\alpha \in [0,1]$, then $A(x_1) \geq \lambda$ and $A(x_2) \geq \lambda$. It follows from the convexity of A that $A(\alpha x_1 + (1-\alpha) x_2) \geq \min(A(x_1), A(x_2)) \geq \lambda$, i.e. $\alpha x_1 + (1-\alpha) x_2 \in A_\lambda$, and thus A_λ is a convex set.

Sufficiency. Assume that A_λ is a convex set for every $\lambda \in [0,1]$. Especially, for $x_1, x_2 \in X$, A_λ is convex for $\lambda = A(x_1) \wedge A(x_2)$. Since $A(x_1) \geq \lambda$ and $A(x_2) \geq \lambda$, we have $x_1 \in A_\lambda$ and $x_2 \in A_\lambda$, whence $\alpha x_1 + (1-\alpha) x_2 \in A_\lambda$. Therefore, $A(\alpha x_1 + (1-\alpha) x_2) \geq \lambda = A(x_1) \wedge A(x_2)$, which indicates A is a convex fuzzy set on X. □

Proposition 4.7. *The intersection $A_1 \cap A_2$ of two convex fuzzy sets A_1 and A_2 on X is a convex fuzzy set on X.*

Proof. By the convexity of A_1 and A_2, for $x_1, x_2 \in X$ and $\alpha \in [0,1]$,

$$(A_1 \cap A_2)(\alpha x_1 + (1-\alpha)x_2) = A_1(\alpha x_1 + (1-\alpha)x_2) \wedge A_2(\alpha x_1 + (1-\alpha)x_2)$$
$$\geq (A_1(x_1) \wedge A_1(x_2)) \wedge (A_2(x_1) \wedge A_2(x_2))$$
$$= (A_1 \cap A_2)(x_1) \wedge (A_1 \cap A_2)(x_2),$$

which implies the convexity of $A_1 \cap A_2$. □

Definition 4.5. *If a fuzzy quantity is convex, then it is called a convex fuzzy quantity.*

Clearly, a fuzzy quantity A is convex iff

$$\forall x, x_1, x_2 \in \mathbb{R}, x_1 < x < x_2, A(x) \geq A(x_1) \wedge A(x_2).$$

Noticing that every non-empty convex set on the real line is an interval, every α-cut of a convex fuzzy quantity is an interval or the empty set by Proposition 4.6.

Theorem 4.3. *A fuzzy quantity A is convex iff at least one of the following statements is true:*

(1) A is increasing;
(2) A is decreasing;
(3) There exists an $x_0 \in \mathbb{R}$ such that A is increasing in $]-\infty, x_0[$ and decreasing in $]x_0, +\infty[$ and

$$\forall x_1 \in]-\infty, x_0[, x_2 \in]x_0, +\infty[, A(x_0) \geq A(x_1) \wedge A(x_2).$$

Proof. Sufficiency. Let $x_1 < x < x_2$.

If (1) is true, then $A(x) \geq A(x_1) \geq A(x_1) \wedge A(x_2)$.

If (2) is true, then $A(x) \geq A(x_2) \geq A(x_1) \wedge A(x_2)$. Hence A is convex if A is increasing or decreasing.

Now we prove that $A(x) \geq A(x_1) \wedge A(x_2)$ when (3) is true.

If $x_1 \geq x_0$ or $x_2 \leq x_0$, the proof is similar to that under (1) or (2) respectively.

If $x_1 < x_0 < x_2$, consider the three cases: (i) $x < x_0$; (ii) $x > x_0$; (iii) $x = x_0$.

In the case (i), $A(x) \geq A(x_1) \geq A(x_1) \wedge A(x_2)$;
In the case (ii), $A(x) \geq A(x_2) \geq A(x_1) \wedge A(x_2)$;
In the case (iii), $A(x) \geq A(x_1) \wedge A(x_2)$ is just a given condition.

In summary, A is a convex fuzzy quantity whatever.

Necessity. Assume that A is a convex fuzzy quantity. Write $z_0 = \text{hgt}(A)$.

4.3 Convex Fuzzy Quantities

Case I. If z_0 is attained, i.e. there exists an x_0 such that $A(x_0) = z_0$, then $A(x_2) \geq A(x_1) \wedge A(x_0) = A(x_1)$ for $x_1 < x_2 < x_0$. As a result, A is increasing in $]-\infty, x_0[$. Similarly, A is decreasing in $]x_0, +\infty[$. $A(x_0) \geq A(x_1)) \wedge A(x_2)$ ($\forall x_1 \in]-\infty, x_0[, x_2 \in]x_0, +\infty[$) follows from the convexity of A. In this case, statement (3) is true.

Case II. If z_0 is not attained, then $A(x) < z_0$ for $x \in \mathbb{R}$. In this case, there exists a strictly increasing sequence $\{A(x_n)\}$ such that $A(x_n) \to z_0$ as $n \to +\infty$.

If an accumulation point x_0 of $\{x_n\}$ can be found, then assume that $x_{n_k} \to x_0 (k \to +\infty)$. For $x_1 < x_2 < x_0$, there exists k such that $A(x_{n_k}) > A(x_1)$ and $x_1 < x_2 < x_{n_k}$. Then $A(x_2) \geq A(x_{n_k}) \wedge A(x_1) = A(x_1)$, that is, A is increasing in $]-\infty, x_0[$. A similar reasoning may lead to the conclusion that A is decreasing in $]x_0, +\infty[$. The inequality $A(x_0) \geq A(x_1)) \wedge A(x_2)$ ($\forall x_1 \in]-\infty, x_0[, x_2 \in]x_0, +\infty[$) follows from the convexity of A again. Therefore, statement (3) is true.

If $\{x_n\}$ has no accumulation point, then there exists $\{x_{n_k}\}$, $x_{n_k} \to +\infty$ or $x_{n_k} \to -\infty$ as $k \to +\infty$. In the case $x_{n_k} \to +\infty$, for $x_1 < x_2$, we can find k, which is sufficiently large, such that $x_1 < x_2 < x_{n_k}$ and $A(x_{n_k}) > A(x_1)$. Applying the convexity of A yields $A(x_2) \geq A(x_1)$, i.e. A is increasing in $\mathbb{R}$, and thus statement (1) is true. In the case $x_{n_k} \to -\infty$, a similar proof leads to the conclusion that A is decreasing and thus statement (2) is true. □

The above theorem shows that the membership function of a convex fuzzy quantity is unimodal in its general case. In the rest of this section, we turn to the mathematical representation of convex fuzzy quantities.

Definition 4.6. *Let $I(\mathbb{R})$ be the set of all intervals and the empty set on the real line. If a mapping $I : [0, 1] \to I(\mathbb{R})$ is a nest of sets, then I is called a nest of intervals.*

Theorem 4.4. *If I is a nest of intervals, then the fuzzy quantity $A = \bigcup_{\alpha \in [0,1]} \alpha I(\alpha)$ is a convex fuzzy quantity.*

Proof. By the Representation Theorem, $A_\alpha = \bigcap_{\lambda < \alpha} I(\lambda)$. Since every $I(\lambda)$ is an interval, A_α is an interval or the empty set for all $\alpha \in [0, 1]$. Hence A is a convex fuzzy quantity. □

Example 4.6. Let

$$I(\alpha) = \begin{cases} [5\alpha - 6, 4 - 10\alpha] & 0 \leq \alpha \leq 2/3, \\ \emptyset & 2/3 < \alpha \leq 1. \end{cases}$$

Then I is a nest of intervals and the convex fuzzy quantity determined by I is

$$A(x) = \bigvee_{\alpha \in [0,1]} \alpha \wedge I(\alpha)(x) = \bigvee_{\alpha \in [0, 2/3]} \{\alpha | x \in [5\alpha - 6, 4 - 10\alpha]\}.$$

If $x < -6$ or $x > 4$, then $A(x) = 0$.

If $-6 \leq x \leq 4$, we have $A(x) = \bigvee_{\alpha \in [0,1]} \{\alpha | \alpha \leq \min\{\frac{x+6}{5}, \frac{4-x}{10}\}\}$.

Hence $A(x) = \frac{x+6}{5}$ or $\frac{4-x}{10}$ according as $x \leq -\frac{8}{3}$ or $x > -\frac{8}{3}$.

In summary,
$$A(x) = \begin{cases} \frac{x+6}{5} & -6 \leq x \leq -\frac{8}{3} \\ \frac{4-x}{10} & -\frac{8}{3} < x < 4 \\ 0 & otherwise. \end{cases}$$

4.4 Fuzzy Numbers

4.4.1 The Concept of a Fuzzy Number

Definition 4.7. *If a fuzzy quantity A satisfies that for every $\alpha \in]0,1]$, A_α is a closed interval, then A is called a fuzzy number.*

Since A_1 is a closed interval for a fuzzy number A, A is normal. In addition, every fuzzy number is a convex fuzzy quantity by Proposition 4.6.

Theorem 4.5. $A \in F(\mathbb{R})$ *is a fuzzy number iff there exist $a, b \in \mathbb{R}$ such that*
$$A(x) = \begin{cases} 1 & x \in [a, b], \\ L(x) & x < a, \\ R(x) & x > b, \end{cases}$$
where L is increasing and right continuous in $]-\infty, a[$ and $L(x) \to 0$ as $x \to -\infty$, and R is decreasing and left continuous in $]b, +\infty[$ and $R(x) \to 0$ as $x \to +\infty$.

Proof. Necessity. If A is a fuzzy number, then $A_1 = \ker(A) \neq \emptyset$. We assume that $A_1 = [a, b]$.

When $x \in [a, b]$, it is obvious that $A(x) = 1$.

When $x < a$, let $L(x) = A(x)$. We proceed to prove that L satisfies the desired requirements.

Firstly, it follows from the convexity of A that for $-\infty < x_1 < x_2 < a$, $A(x_2) \geq A(x_1) \wedge A(a) = A(x_1)$, i.e. $L(x_2) \geq L(x_1)$, which indicates that L is increasing in $]-\infty, a[$.

Secondly, we show that L is right continuous in $]-\infty, a[$ by contradiction. Suppose that L is not right continuous for some $x_0 < a$. Then there exists $x_n \to x_0$ such that $x_n > x_0$ and $\lim_{n \to +\infty} L(x_n) \neq L(x_0)$. Since L is increasing, $\lim_{n \to +\infty} L(x_n) > L(x_0)$. Without loss of generality, assume that $\{x_n\}$ is decreasing and $x_n < a$. Write $\lim_{n \to +\infty} L(x_n) = \alpha$. From $A(x_0) = L(x_0) < \alpha$,

4.4 Fuzzy Numbers

we know that $x_0 \notin A_\alpha$. However $A(x_n) = L(x_n) \geq \alpha$ implies $x_n \in A_\alpha$, which contradicts the closedness of A_α.

What is left is the proof of the equality $\lim_{x \to -\infty} L(x) = 0$.

Otherwise, $\lim_{x \to -\infty} L(x) = \beta > 0$. It means that $\forall x < a$, $A(x) = L(x) \geq \beta$. In other words, $x \in A_\beta$ holds for $x < a$, which implies that A_β is an unbounded set. It is a contradiction with the definition of a fuzzy number.

When $x > b$, let $R(x) = A(x)$. Following a similar reasoning, we can show that R is decreasing and left continuous in $]b, +\infty[$ and $R(x) \to 0$ as $x \to +\infty$.

Sufficiency. For $\alpha \in]0, 1]$, write

$$a_\alpha = \inf\{x | x < a, L(x) \geq \alpha\} \text{ and } b_\alpha = \sup\{x | x > a, R(x) \geq \alpha\}.$$

The restriction $a_\alpha \leq a$, together with $\lim_{x \to -\infty} L(x) = 0$, implies that a_α ($\alpha \in]0, 1]$) is a finite number. Similarly b_α ($\alpha \in]0, 1]$) is also a finite number. To prove that A is a fuzzy number, it suffices to prove that $A_\alpha = [a_\alpha, b_\alpha]$.

If $x \notin [a_\alpha, b_\alpha]$, then $x < a_\alpha$ or $x > b_\alpha$, either of which leads to $A(x) < \alpha$, i.e. $x \notin A_\alpha$. This proves that $A_\alpha \subseteq [a_\alpha, b_\alpha]$.

Conversely, if $x \in [a_\alpha, b_\alpha] = [a_\alpha, a[\cup[a, b]\cup]b, b_\alpha]$, consider the cases listed below. We prove that $x \in A_\alpha$ in each case.

Case (i). If $x \in [a, b]$, then $A(x) = 1$. Hence $x \in A_\alpha$ for every $\alpha \in]0, 1]$.
Case (ii). If $x \in]a_\alpha, a[$, then $a > x > a_\alpha = \inf\{x | x < a, L(x) \geq \alpha\}$. There exists an x_0 such that $x_0 < x$ and $L(x_0) \geq \alpha$. Therefore $L(x) \geq L(x_0) \geq \alpha$, i.e. $A(x) \geq \alpha$, or equivalently $x \in A_\alpha$.
Case (iii). If $x \in]b, b_\alpha[$, the proof is similar to that in the case (ii).
Case (iv). If $x = a_\alpha < a$, $A(x) = A(a_\alpha) = L(a_\alpha) = \lim_{x \to a_\alpha^+} L(x) \geq \alpha$ due to the right continuity of L and the case (ii).
Case (v). If $x = b_\alpha > b$, the proof is similar to the one in the case (iv).

Hence, $[a_\alpha, b_\alpha] \subseteq A_\alpha$, and the two sets are identical. □

Figure 4.1 depicts a fuzzy number.

Fig. 4.1 A general form of a fuzzy number

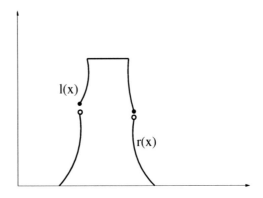

From now on, we shall call L and R the left spread and right spread of the fuzzy number A respectively.

Based on Theorem 4.5, we list some extensively used fuzzy numbers.

(1) $A = 1/a$ $(a \in \mathbb{R})$ is a fuzzy number, i.e. every real number is a fuzzy number;

(2) The closed interval $A = [a, b]$ is a fuzzy number since the membership function of A is
$$A(x) = \begin{cases} 1 & x \in [a, b] \\ 0 & \text{otherwise} \end{cases}.$$

(3) If $A(x) = \begin{cases} \dfrac{x-a}{b-a} & x \in [a, b] \\ \dfrac{x-c}{b-c} & x \in [b, c] \\ 0 & \text{otherwise} \end{cases}$, then A is a fuzzy number, which is called a triangular fuzzy number, denoted (a, b, c). For instance, $(1, 2, 3)$ is a triangular fuzzy number, which may be used to describe the fuzzy notion of *around 2*.

(4) If $A(x) = \begin{cases} \dfrac{x-a}{b-a} & x \in [a, b] \\ 1 & x \in [b, c] \\ \dfrac{x-d}{c-d} & x \in [c, d] \\ 0 & \text{otherwise} \end{cases}$, then A is a fuzzy number which is called a trapezoidal fuzzy number, denoted (a, b, c, d). For instance, $(1, 2, 3, 4)$ is a trapezoidal fuzzy number, which may be used to describe the notion of *approximately between 2 and 3*. Clearly, a triangular fuzzy number is a special trapezoidal fuzzy number with a single point in the kernel.

(5) For $\alpha, \beta, \gamma \in \mathbb{R}$ such that $\alpha < \beta < \gamma$ and $\beta = \frac{\alpha+\gamma}{2}$, define
$$S(x; \alpha, \beta, \gamma) = \begin{cases} 0 & x \in]-\infty, \alpha[\\ 2(\frac{x-\alpha}{\gamma-\alpha})^2 & x \in [\alpha, \beta] \\ 1 - 2(\frac{x-\gamma}{\gamma-\alpha})^2 & x \in [\beta, \gamma] \\ 1 & x \in [\gamma, +\infty[\end{cases},$$

which is called the S-membership function. Clearly, $S(\cdot; \alpha, \beta, \gamma)$ is a fuzzy number for any $\alpha, \beta, \gamma \in \mathbb{R}$. This type of fuzzy numbers may be used to represent increasing notions such as old, tall, and high. The complement of an S-membership function is useful to represent decreasing notions such as young, small, and low. A fuzzy notion such as old for the age of a person may be represented by means of the membership function $S(.; 50, 60, 70)$. Similarly, a high blood pressure for an adult may be represented as $S(.; 14, 15, 16)$.

(6) By means of the S-membership function, we define another type of fuzzy numbers which is called the π-membership function by: for β, $\beta > 0$, and $\gamma, \gamma \in \mathbb{R}$,

4.4 Fuzzy Numbers

$$\pi(x;\beta,\gamma) = \begin{cases} S(x;\gamma-\beta,\gamma-\frac{\beta}{2},\gamma) & x \in]-\infty,\gamma[\\ 1 - S(x;\gamma,\gamma+\frac{\beta}{2},\gamma+\beta) & x \in [\gamma,+\infty[. \end{cases}$$

This type of fuzzy numbers can be used to represent fuzzy notions such as approximately 2 (years), about 140 (pounds), etc.

Definition 4.8. *Denote the set of closed intervals by $CI(\mathbb{R})$. If a mapping I from $[0,1]$ to $CI(\mathbb{R})$ satisfies that I is a nest of sets, then I is called a nest of closed intervals.*

Theorem 4.6. *(Representation Theorem of Fuzzy Numbers) Let I be a nest of closed intervals and $A = \bigcup_{\alpha \in [0,1]} \alpha I(\alpha)$. If $I(\alpha) = [i_1(\alpha), i_2(\alpha)]$, then*

(1) $\forall \alpha \in (0,1]$, $A_\alpha = [\lim_{\lambda \to \alpha^-} i_1(\lambda), \lim_{\lambda \to \alpha^-} i_2(\lambda)]$;
(2) A is a fuzzy number with its membership function

$$A(x) = \begin{cases} 1 & x \in [i_1(1), i_2(1)] \\ L(x) & x < i_1(1) \\ R(x) & x > i_2(1) \end{cases}$$

where $L(x) = \bigvee_{\alpha \in [0,1]} \{\alpha | i_1(\alpha) \leq x\}$, $R(x) = \bigvee_{\alpha \in [0,1]} \{\alpha | i_2(\alpha) \geq x\}$.

Proof. (1) Applying the Representation Theorem yields

$$A_\alpha = \bigcap_{\lambda < \alpha} I(\lambda) = \bigcap_{\lambda < \alpha} [i_1(\lambda), i_2(\lambda)]$$
$$= [\bigvee_{\lambda < \alpha} i_1(\lambda), \bigwedge_{\lambda < \alpha} i_2(\lambda)] = [\lim_{\lambda \to \alpha^-} i_1(\lambda), \lim_{\lambda \to \alpha^-} i_2(\lambda)].$$

(2) It follows from (1) that A is a fuzzy number. Furthermore,

$$A(x) = \bigvee_{\alpha \in [0,1]} \alpha \wedge [i_1(\alpha), i_2(\alpha)](x) = \bigvee_{\alpha \in [0,1]} \{\alpha | i_1(\alpha) \leq x \leq i_2(\alpha)\}$$

$$= \begin{cases} 1 & x \in [i_1(1), i_2(1)] \\ \bigvee_{\alpha \in [0,1]} \{\alpha | i_1(\alpha) \leq x\} & x < i_1(1) \\ \bigvee_{\alpha \in [0,1]} \{\alpha | i_2(\alpha) \geq x\} & x > i_2(1) \end{cases}$$

$$= \begin{cases} 1 & x \in [i_1(1), i_2(1)] \\ L(x) & x < i_1(1) \\ R(x) & x > i_2(1) \end{cases}. \qquad \Box$$

Example 4.7. *If $I(\alpha) = [\alpha, 2-\alpha]$, then I is a nest of closed intervals with $i_1(\alpha) = \alpha$, $i_2(\alpha) = 2 - \alpha$. Thus $i_1(1) = i_2(1) = 1$. Consequently,*

$$L(x) = \bigvee_{\alpha \in [0,1]} \{\alpha | i_1(\alpha) \leq x\} = x,$$

$$R(x) = \bigvee_{\alpha \in [0,1]} \{\alpha | i_2(\alpha) \geq x\} = 2 - x.$$

The fuzzy number determined by I is

$$A(x) = \begin{cases} x & 0 \leq x \leq 1 \\ 2 - x & 1 < x \leq 2 \\ 0 & \text{elsewhere} \end{cases},$$

i.e. $A = (0, 1, 2)$.

4.4.2 Properties of Algebraic Operations on Fuzzy Numbers

Compared with fuzzy quantities, fuzzy numbers show a desired behavior as will be seen in this section.

Proposition 4.8. *If a real function f is continuous and if A is a fuzzy number, then $(f(A))_\alpha = f(A_\alpha)$ for every $\alpha \in]0,1]$.*

Proof. By Proposition 4.3(4), $f(A_\alpha) \subseteq (f(A))_\alpha$. To prove the conclusion, it suffices to show the reverse inclusion is true.

If $y \in (f(A))_\alpha$, then $(f(A))(y) \geq \alpha$, i.e. $\beta = \bigvee_{f(x)=y} A(x) \geq \alpha > 0$. Write $\alpha_n = \beta - 1/n$. We assume that $\alpha_{n_0} > 0$. Then a sequence $\{x_n\}$ can be found such that $f(x_n) = y$ and $A(x_n) > \alpha_n$. As $n > n_0$, $x_n \in A_{\alpha_n} \subseteq A_{\alpha_{n_0}}$, and thus $\{x_n\}$ is bounded. Take a convergent subsequence $\{x_{n_k}\}$ of $\{x_n\}$ with the limit x_0. From $A(x_{n_k}) > \alpha_{n_k} = \beta - 1/n_k$, it follows that $A(x_{n_k}) > \beta - 1/N$ for any fixed N if k is sufficiently large. In other words, $x_{n_k} \in A_{\alpha_N}$ for any fixed N as long as k is large enough. Since A_{α_N} is closed, $x_0 \in A_{\alpha_N}$, i.e. $A(x_0) \geq \alpha_N$. Now the arbitrariness of N implies $A(x_0) \geq \beta \geq \alpha$, thus $x_0 \in A_\alpha$. From the continuity of f and the equality $f(x_{n_k}) = y$, we have $f(x_0) = y$. It indicates that $y = f(x_0) \in f(A_\alpha)$, that is to say, $(f(A))_\alpha \subseteq f(A_\alpha)$. □

Using the same reasoning as in Proposition 4.8, the conclusion can be generalized.

Proposition 4.9. *If a real function f of n real variables is continuous and if $A_1, A_2, \cdots, A_n$ are fuzzy numbers, then*

$$(f(A_1,, A_2, \cdots, A_n))_\alpha = f((A_1)_\alpha, (A_2)_\alpha, \cdots, (A_n)_\alpha)$$

for all $\alpha \in (0, 1]$.

4.4 Fuzzy Numbers

Applying Proposition 4.9, many properties of fuzzy numbers in terms of α-cut can be derived. we list some of them in the following. Let A, B be fuzzy numbers. Write $A_\alpha = [a_\alpha^-, a_\alpha^+]$ and $B_\alpha = [b_\alpha^-, b_\alpha^+]$.

(1) $(A+B)_\alpha = [a_\alpha^- + b_\alpha^-, a_\alpha^+ + b_\alpha^+]$.
(2) $(A-B)_\alpha = [a_\alpha^- - b_\alpha^+, a_\alpha^+ - b_\alpha^-]$.
(3) $(AB)_\alpha = [\min(a_\alpha^- b_\alpha^-, a_\alpha^- b_\alpha^+, a_\alpha^+ b_\alpha^-, a_\alpha^+ b_\alpha^+),$

$$\max(a_\alpha^- b_\alpha^-, a_\alpha^- b_\alpha^+, a_\alpha^+ b_\alpha^-, a_\alpha^+ b_\alpha^+)].$$

(4) If $0 \notin \mathrm{supp}(B)$, then

$$(A/B)_\alpha = [\min(a_\alpha^-/b_\alpha^-, a_\alpha^-/b_\alpha^+, a_\alpha^+/b_\alpha^-, a_\alpha^+/b_\alpha^+),$$
$$\max(a_\alpha^-/b_\alpha^-, a_\alpha^-/b_\alpha^+, a_\alpha^+/b_\alpha^-, a_\alpha^+/b_\alpha^+)].$$

(5) $(A \vee B)_\alpha = [a_\alpha^- \vee b_\alpha^-, a_\alpha^+ \vee b_\alpha^+]$.
(6) $(A \wedge B)_\alpha = [a_\alpha^- \wedge b_\alpha^-, a_\alpha^+ \wedge b_\alpha^+]$.

These properties are direct results of Proposition 4.9 and Example 4.4. For instance,

$$(A+B)_\alpha = A_\alpha + B_\alpha = [a_\alpha^-, a_\alpha^+] + [b_\alpha^-, b_\alpha^+] = [a_\alpha^- + b_\alpha^-, a_\alpha^+ + b_\alpha^+].$$

From these properties, we know that the sum, difference, product, quotient, maximum, minimum of two fuzzy numbers are still fuzzy numbers. In addition, further properties can be deduced from these known ones.

(7) $A \wedge (A \vee B) = A$, $\quad A \vee (A \wedge B) = A$.
(8) $A \wedge (B \vee C) = (A \wedge B) \vee (A \wedge C)$, $\quad A \vee (B \wedge C) = (A \vee B) \wedge (A \vee C)$.
(9) $(A \vee B) + (A \wedge B) = A + B$.
(10) $A \vee B = A \Leftrightarrow A \wedge B = B$.

Using α-cuts, these identities are easily verified. For instance, let $A_\alpha = [a_\alpha^-, a_\alpha^+]$ and $B_\alpha = [b_\alpha^-, b_\alpha^+]$. Then

$$[(A \vee B) + (A \wedge B)]_\alpha = (A \vee B)_\alpha + (A \wedge B)_\alpha$$
$$= [a_\alpha^- \vee b_\alpha^-, a_\alpha^+ \vee b_\alpha^+] + [a_\alpha^- \wedge b_\alpha^-, a_\alpha^+ \wedge b_\alpha^+]$$
$$= [a_\alpha^- \vee b_\alpha^- + a_\alpha^- \wedge b_\alpha^-, a_\alpha^+ \vee b_\alpha^+ + a_\alpha^+ \wedge b_\alpha^+]$$
$$= [a_\alpha^- + b_\alpha^-, a_\alpha^+ + b_\alpha^+] = (A+B)_\alpha,$$

which proves the validity of equality (9).

For trapezoidal or triangular fuzzy numbers, some special formulae for algebraic operations are available.

Let $A - (a_1, b_1, c_1, d_1)$ and $D - (a_2, b_2, c_2, d_2)$.

(11) $A + B = (a_1 + a_2, b_1 + b_2, c_1 + c_2, d_1 + d_2)$;
(12) $A - B = (a_1 - d_2, b_1 - c_2, c_1 - b_2, d_1 - a_2)$;

We prove the validity of equality (11) as an example. Simple calculations yield
$$A_\alpha = [a_1 + (b_1 - a_1)\alpha, d_1 + (c_1 - d_1)\alpha]$$
$$B_\alpha = [a_2 + (b_2 - a_2)\alpha, d_2 + (c_2 - d_2)\alpha].$$
Let $C = (a_1 + a_2, b_1 + b_2, c_1 + c_2, d_1 + d_2)$. Then
$$C_\alpha = [(a_1 + a_2) + (b_1 + b_2 - a_1 - a_2)\alpha, (d_1 + d_2) + (c_1 + c_2 - d_1 - d_2)\alpha],$$
which indicates that $C_\alpha = A_\alpha + B_\alpha = (A + B)_\alpha$ for every $\alpha \in [0, 1]$. Thus $A + B = C = (a_1 + a_2, b_1 + b_2, c_1 + c_2, d_1 + d_2)$.

The proof of the remaining identity is similar. □

Calculations associated with general algebraic operations of fuzzy numbers are tedious even if the involved fuzzy numbers are trapezoidal. However, it can be seen from the preceding properties that algebraic operations of their cuts are well behaved. We can obtain resulting fuzzy numbers by performing operations of cuts of fuzzy numbers and applying Decomposition Theorem I afterwards. Here is an example.

Example 4.8. Let $A = (0, 1, 2)$ and $B = (2, 3, 4)$. Then $A_\alpha = [\alpha, 2 - \alpha]$ and $B_\alpha = [2 + \alpha, 4 - \alpha]$. Denote $A_\alpha B_\alpha = [2\alpha + \alpha^2, 8 - 6\alpha + \alpha^2]$ by $I(\alpha) = [i_1(\alpha), i_2(\alpha)]$.

$$AB = \bigcup_{\alpha \in [0,1]} (AB)_\alpha = \bigcup_{\alpha \in [0,1]} A_\alpha B_\alpha = \bigcup_{\alpha \in [0,1]} \alpha I(\alpha).$$

Since $i_1(\alpha) = 2\alpha + \alpha^2$ and $i_2(\alpha) = 8 - 6\alpha + \alpha^2$, we have $I(1) = [i_1(1), i_2(1)] = 3$. When $x < 3$, $L(x) = \bigvee_{\alpha \in [0,1]} \{\alpha | i_1(\alpha) \le x\} = \bigvee_{\alpha \in [0,1]} \{\alpha | \alpha + \alpha^2 \le x\} = 0 \vee (\sqrt{1+x} - 1)$.

Similarly, $R(x) = 0 \vee (3 - \sqrt{1+x})$ when $x > 3$.

We write it in one formula,

$$(AB)(x) = \begin{cases} \sqrt{1+x} - 1 & 0 \le x \le 3 \\ 3 - \sqrt{1+x} & 3 < x \le 8 \\ 0 & \text{otherwise.} \end{cases}$$

4.5 Ranking of Fuzzy Numbers

Currently, almost every ranking approach is concerned with one or multiple ranking indices. As a result, more than forty ranking indices can be found in the literature. In order to investigate the ranking problem in detail, let us have a general look at these indices.

Assume that $\mathcal{A} = \{A_1, A_2, \cdots, A_n\}$ is the set of fuzzy numbers to be ranked. Roughly speaking, all the ranking indices fall into three categories.

4.5 Ranking of Fuzzy Numbers

In the first category, a mapping $F : \mathcal{A} \to \mathbb{R}$, called a ranking function, is constructed to transform every A_i into a real number $F(A_i)$, which serves as the ranking index, and then simply ranks $A_1, A_2, \ldots, A_n$ according to the natural order of real numbers $F(A_1), F(A_2), \cdots, F(A_n)$. In the second category, one defines some certain reference set(s) and then evaluates A_i by comparing A_i to the reference set(s) to form the ranking index. In the last category, a fuzzy relation R on $\mathcal{A} \times \mathcal{A}$, which serves as the ranking index, is constructed to make pairwise comparison between A_i and A_j. The final ranking of $A_1, A_2, \ldots, A_n$ is obtained through some specific procedure if more than two fuzzy numbers are involved.

To take a decision maker's risk attitude into account, many authors introduce a parameter in a ranking index to adjust the ranking approach to suit different type of decision makers. Considering that any single index merely reflects a partial point of view on fuzzy numbers, some authors suggest the joint use of multiple indices simultaneously. In the following, we introduce some typical ranking indices in each category. To ensure the applicability of some ranking indices, we assume that the involved fuzzy numbers are limited, i.e. their supports are bounded in this section. For two fuzzy numbers A and B, we use $A \succ B$ and $A \sim B$ to denote "A has a higher rank than B" and "A and B have the same rank" respectively, and the notation $A \succsim B$ denotes $A \succ B$ or $A \sim B$.

4.5.1 Ranking Fuzzy Numbers by a Ranking Function

The first authors who use this type of approaches are Adamo [2] and Yager [164]. Adamo fixes an $\alpha \in]0,1]$ and simply evaluates a fuzzy number by using the right endpoint of its α-cut interval, i.e. a larger right endpoint corresponds to a higher rank. More formally, let A be a fuzzy number with $A_\alpha = [a_\alpha, b_\alpha]$. Then Adamo uses the ranking function $AD^\alpha(A) = b_\alpha$. A larger $AD^\alpha(A)$ corresponds to a higher rank. Clearly, this method is heavily dependent on the choice of the α and ignores the left spread of a fuzzy number. Yager also proposes a ranking function which is based on α-cut. What is the difference between them is that Yager uses all the α-cuts simultaneously. Yager's evaluating index for a fuzzy number A with α-cut $A_\alpha = [a_\alpha, b_\alpha]$ is

$$Y(A) = \frac{1}{2}\int_0^1 (a_\alpha + b_\alpha)d\alpha.$$

Yager's method tends to be neutral by considering the left and right endpoint of α-cut equally. To reflect a decision maker's risk type, the following general index may be considered

$$Y^\lambda(A) = \frac{1}{2}\int_0^1 ((1-\lambda)a_\alpha + \lambda b_\alpha)d\alpha,$$

where $\lambda \in [0,1]$ represents the decision maker's risk attitude. The larger the λ is, the more risky the decision maker tends to be.
When $\lambda = 0$,
$$Y^0(A) = \frac{1}{2}\int_0^1 a_\alpha d\alpha;$$
When $\lambda = 1$,
$$Y^1(A) = \frac{1}{2}\int_0^1 b_\alpha d\alpha;$$
When $\lambda = 0.5$,
$$Y^{0.5}(A) = \frac{1}{2}\int_0^1 (a_\alpha + b_\alpha)d\alpha = Y(A).$$

The above mentioned indices are related to α-cuts of the fuzzy numbers in question. In applying them, it is not necessary to have the exact expression of the membership functions of the involved fuzzy numbers. This is very useful in practice. As known to us, the function or arithmetical operations of fuzzy numbers defined by the Zadeh's extension principle is very complicated in calculation. As a result, the exact expression of fuzzy numbers is often difficult to acquire. However, the corresponding operations of α-cuts are merely concerned with the two endpoints of α-cuts, which are relatively easier in the implementation.

Besides these indices based on α-cut, there are some others which can be classified into this category [22, 26, 59, 89, 165], among which we mention the centroid index of a fuzzy number, defined by

$$C(A) = \frac{\int_{supp(A)} xA(x)dx}{\int_{supp(A)} A(x)dx}.$$

A larger centroid index corresponds to a higher rank.

Example 4.9. *Let $A = (1, 2, 3)$ and $B = (0, 2, 4)$. Then $A_\alpha = [1 + \alpha, 3 - \alpha]$ and $B_\alpha = [2\alpha, 4 - 2\alpha]$. Hence $AD^\alpha(A) = 3 - \alpha$ and $AD^\alpha(B) = 4 - 2\alpha$. By the Adamo's method, $B \succ A$ for all $\alpha \in]0, 1[$.*

In addition, $Y^\lambda(A) = \lambda + 1.5$ and $Y^\lambda(B) = 2\lambda + 1$. Consequently, $A \sim B$, $A \succ B$ and $B \succ A$ by $Y^{0.5}$, Y^0 and Y^1 respectively.

Finally, $C(A) = C(B) = 2$, and hence $A \sim B$ by the centroid index.

4.5.2 Ranking Fuzzy Numbers According to the Closeness to a Reference Set

For this class of methods, one or more reference sets are defined and then the ranking index is constructed by comparing the fuzzy number to be ranked to the reference(s). The distinct definitions of reference set(s) may be adopted by different authors. The most frequently used one is the fuzzy maximum

4.5 Ranking of Fuzzy Numbers

(fuzzy minimum) defined by the Zadeh's extension principle. Let us have a closer look at this type of ranking methods.

Assume that $A_1, A_2, \ldots, A_n$ are n fuzzy numbers to be ranked. According to the Zadeh's extension principle, their fuzzy maximum $\widetilde{max}$ is defined by

$$\widetilde{max}\{A_1, A_2, \ldots, A_n\}(x) = \sup_{\max(x_1,\ldots,x_n)=x} \bigwedge_{i=1}^{n} A_i(x_i).$$

The fuzzy maximum of n fuzzy numbers is still a fuzzy number. However, unlike in the case of real numbers, $\widetilde{max}\{A_1, A_2, \ldots, A_n\}$ is not generally one of $A_1, A_2, \ldots, A_n$. Therefore, the fuzzy $\widetilde{max}$ is merely used as a reference set for the ranking of fuzzy numbers. Noticing the insufficiency of fuzzy $\widetilde{max}$, Kerre[75] calculates the Hamming distance between $\widetilde{max}(A_1, A_2, \ldots, A_n)$ and A_i ($i = 1, 2, \ldots, n$) to determine the rank of A_i. Therefore, the Kerre's index is

$$K(A_i) = \int_S |A_i(x) - \widetilde{max}(A_1, A_2, \ldots, A_n)(x)| dx,$$

where $S = \bigcup_{i=1}^{n} supp A_i$. The resulting order relation is formulated by:

$$A_i \succ A_j \Leftrightarrow K(A_i) < K(A_j);$$

$$A_i \sim A_j \Leftrightarrow K(A_i) = K(A_j).$$

Clearly, $K(A_i)$ ranks fuzzy number A_i by considering the closeness of A_i to the fuzzy maximum (reference set) in the sense of the Hamming distance. The Kerre's method can be generalized by replacing the Hamming distance by a nearness measure $N[64]$, i.e. we may define the ranking index

$$W^N(A_i) = N(A_i, \widetilde{max}\{A_1, A_2, \cdots, A_n\})(i = 1, 2, \cdots, n).$$

A larger $W^N(A_i)$ corresponds to a higher ranking.

For instance, if N_{M_1} is selected, then the ranking index is

$$W^{N_{M_1}}(A_i) = \frac{\int_{\mathbb{R}} (A_i(x) \wedge \widetilde{max}\{A_1, A_2, \cdots, A_n\}(x)) dx}{\int_{\mathbb{R}} (A_i(x) \vee \widetilde{max}\{A_1, A_2, \cdots, A_n\}(x)) dx}.$$

Instead of the fuzzy maximum, other reference sets are also visible in the literature.

For example, Chen[27] defines the fuzzy maximizing set by

$$A_{max}^k(x) = \left(\frac{x - x_{min}}{x_{max} - x_{min}}\right)^k (x \in [x_{min}, x_{max}]),$$

where $x_{\max} = \sup \bigcup_{i=1}^{n} \operatorname{supp} A_i$, $x_{\min} = \inf \bigcup_{i=1}^{n} \operatorname{supp} A_i$ and $k > 0$ a real number.
In order to incorporate the information contained in the left spread of a fuzzy number, the fuzzy minimizing set $A_{\min}^k$, the dual concept of $A_{\max}^k$, is introduced, which is defined by

$$A_{\min}^k(x) = \left(\frac{x_{\max} - x}{x_{\max} - x_{\min}}\right)^k \quad (x \in [x_{\min}, x_{\max}]).$$

Then the left utility $L^k(A_i)$ and the right utility $R^k(A_i)$ of the fuzzy number A_i are defined respectively as

$$L^k(A_i) = \sup \min(A_{\min}^k(x), A_i(x))$$

and

$$R^k(A_i) = \sup \min(A_{\max}^k(x), A_i(x)).$$

Finally, the total utility $T^k(A_i)$ is calculated as

$$T^k(A_i) = \frac{1}{2}(R^k(A_i) + 1 - L^k(A_i)).$$

Then a larger $T^k(A_i)$ corresponds to a higher rank.

Since $L^k(A_i) = N_L(A_i, A_{\min}^k)$ and $R^k(A_i) = N_L(A_i, A_{\max}^k)$, $T^k(A_i)$ evaluates A_i by combining the closeness of A_i to $A_{\max}^k$ and $A_{\min}^k$ (reference sets) in the sense of the lattice nearness measure.

Example 4.10. Let $A = (0.2, 0.3, 1)$, $B = (0, 0.5, 0.8)$. Then $K(A) = \frac{1}{12}$, $K(B) = 0.1$, and thus $A \succ B$ by the Kerre's method.

In addition, $W^{N_{M_1}}(A) = \frac{19}{26}$ and $W^{N_{M_1}}(B) = \frac{7}{9}$, and hence $B \succ A$ by $W^{N_{M_1}}$.

Finally, $T^1(A) = \frac{155}{374}$ and $T^1(B) = \frac{37}{78}$, whence $B \succ A$ by the Chen's method.

Remark 4.1. Besides the foregoing approaches, the ranking indices suggested by Jain [70], Kim and Park [72] also belong to this category.

4.5.3 Ranking Fuzzy Numbers Based on Pairwise Comparisons

In decision analysis, a decision maker is frequently confronted with the comparison of alternatives. Generally, it is difficult to compare all the alternatives directly simultaneously. Instead, comparisons are often made on the pairwise base. To rank fuzzy numbers, some researchers construct fuzzy relations to carry out the pairwise comparisons between them. The following are some typical fuzzy relations on the set $\mathcal{A} = \{A_1, A_2, \cdots, A_n\}$ of fuzzy numbers to be ranked.

4.5 Ranking of Fuzzy Numbers

The first one is based on the Baas and Kwakernaak's ranking index [6]

$$R_{BK}(A_i, A_j) = \sup_{x_i \geq x_j} \min(A_i(x_i), A_j(x_j)).$$

Let $[a_i^-, a_i^+]$ and $[a_j^-, a_j^+]$ be the kernel of A_i and A_j respectively. If there exist $a \in [a_i^-, a_i^+]$ and $b \in [a_j^-, a_j^+]$ such that $a > b$, then $R_{BK}(A_i, A_j) = 1$ and $R_{BK}(A_j, A_i) = hgt(A_i \cap A_j)$ by Exercise 22. Hence, the Baas and Kwakernaak's index compares fuzzy numbers according to the relative position of the kernels and the height of their intersection. Obviously, if $[a_i^-, a_i^+] \cap [a_j^-, a_j^+] \neq \emptyset$, $R_{BK}(A_i, A_j) = R_{BK}(A_j, A_i)) = 1$. The kernel of a fuzzy number represents the scope of its most possible values, whereas the height of the intersection of two fuzzy numbers characterizes the their relative location. Therefore, the Baas and Kwakernaak's index is called very natural in the framework of fuzzy set theory [49], which is employed by many researchers in essence [46, 64, 142].

Another extensively used class of fuzzy relations is constructed by using the so-called upper and lower set and fuzzy maximum or/and fuzzy minimum. For a fuzzy number F, define the upper set $\bar{F}$ and the lower set $\underline{F}$ of F by: $\forall x \in \mathbb{R}$,

$$\bar{F}(x) = \sup_{y \geq x} F(y) \quad \text{and} \quad \underline{F}(x) = \sup_{y \leq x} F(y).$$

Many authors employ the following fuzzy relation to compare A_i and A_j in nature [85, 106, 126, 170],

$$R^\lambda(A_i, A_j) = \lambda d_H(\underline{A}_i, (\underline{A}_i \wedge \underline{A}_j)) + (1 - \lambda) d_H(\bar{A}_i, (\bar{A}_i \wedge \bar{A}_j)),$$

where d_H stands for the Hamming distance and $\lambda \in [0, 1]$ is a risk attitude indicator.

A fuzzy relation can merely provide the information of the pairwise comparisons between A_i and A_j ($i, j = 1, 2, \cdots$). How to determine the final ranking of $A_1, A_2, \cdots, A_n$ by the use of this information is another issue. Now we introduce such a method.

Assume that R on $\mathcal{A} = \{A_1, A_2, \cdots, A_n\}$ is acyclic. Then the following procedure may be used to determine the rank orders of $A_1, A_2, \cdots, A_n$.

Firstly, construct $H_1 = \{A_i \in \mathcal{A} | R(A_i, A_j) \geq R(A_j, A_i) \text{ for } \forall A_j \in \mathcal{A}\}$. Then $H_1 \neq \emptyset$ by Proposition 3.25. If $\mathcal{A}_1 = \mathcal{A} \setminus H_1 \neq \emptyset$, construct $H_2 = \{A_i \in \mathcal{A}_1 | R(A_i, A_j) \geq R(A_j, A_i) \text{ for } \forall A_j \in \mathcal{A}_1\}$. In a similar way, H_3 can be constructed if $\mathcal{A}_2 = \mathcal{A}_1 \setminus H_2 \neq \emptyset$. Keep constructing $H_4, H_5, \cdots, H_m$ until $\mathcal{A}_m = \mathcal{A}_{m-1} \setminus H_m = \emptyset$ for some m. Then the ranking order $\succ$ and $\sim$ are formulated as:

$$A_i \succ A_j \Leftrightarrow \exists H_s, H_t \text{ such that } A_i \in H_s, A_j \in H_t \text{ and } s < t;$$
$$A_i \sim A_j \Leftrightarrow \exists H_s \text{ such that } A_i \in H_s \text{ and } A_j \in H_s.$$

Therefore, the final ranking of $A_1, A_2, \cdots, A_n$ can be determined by the above method if the R on $\mathcal{A}$ satisfies acyclicity.

Remark 4.2. *Acyclicity is a very weak restriction as can be seen from Section 1.6.3 of Chapter 3. For instance, by Exercise 22(2), R_{BK} is weakly transitive, and thus acyclic, and R^λ is consistent, and thus acyclic by Proposition 3.24, whose proof is left as an exercise (Exercise 35). As a matter of fact, it can be known from [150, 151] that almost all fuzzy relations employed for ranking fuzzy quantities satisfy acyclicity if the involved fuzzy quantities are fuzzy numbers.*

Example 4.11. $\mathcal{A} = \{A_1, A_2, A_3, A_4, A_5, A_6\}$

R	A_1	A_2	A_3	A_4	A_5	A_6
A_1	1	0.5	0.3	0.1	0.5	0
A_2	0.5	1	0.7	0.1	0.6	0.2
A_3	0.4	0.9	1	0.2	0.4	0.4
A_4	0.9	0.9	0.7	1	0.4	1
A_5	0.5	0.9	0.9	0.6	1	0.6
A_6	0.9	0.4	0.8	1	0.6	1

It is easily verified that R is weakly transitive and thus acyclic.

For $i = 1, 2, \ldots, 6$, $R(A_6, A_i) \geq R(A_i, A_6)$ and $R(A_5, A_i) \geq R(A_i, A_5)$.

Therefore $H_1 = \{A_5, A_6\}$. Similarly $H_2 = \{A_4\}$, $H_3 = \{A_3\}$, $H_4 = \{A_1, A_2\}$.

The ranking is $A_5 \sim A_6 \succ A_4 \succ A_3 \succ A_1 \sim A_2$.

Proposition 4.10. *The above defined relations $\succ$ and $\sim$ have the properties:*

(1) $\sim$ is reflexive and symmetry.
(2) $\succ$ and $\sim$ are transitive.
(3) Given arbitrary A_i and A_j in $\mathcal{A}$, exactly one of the following relations holds: $A_i \succ A_j$, $A_j \succ A_i$, $A_i \sim A_j$.
(4) $R(A_i, A_j) > R(A_j, A_i) \Rightarrow A_i \succ A_j$.
(5) If R is consistent, then $A_i \succ A_j \Rightarrow R(A_i, A_j) > R(A_j, A_i)$.

Proof. The proof of (1) through (4) is straightforward.

(5) Assume that $A_i \succ A_j$. Then $R(A_i, A_j) \geq R(A_j, A_i)$. Now suppose $R(A_i, A_j) = R(A_j, A_i)$. There must exist $A_k \in \mathcal{A}$ such that

$$R(A_i, A_k) \geq R(A_k, A_i) \text{ and } R(A_j, A_k) < R(A_k, A_j)$$

(Otherwise, $\forall A_k \in \mathcal{A}$, $R(A_j, A_k) \geq R(A_k, A_j)$ if $R(A_i, A_k) \geq R(A_k, A_i)$ and hence $A_j \succsim A_i$). This, together with $R(A_i, A_j) = R(A_j, A_i)$, conflicts with consistency. $\square$

4.5 Ranking of Fuzzy Numbers

It follows immediately from Proposition 4.10(4)(5) that

$$A_i \succ A_j \Leftrightarrow R(A_i, A_j) > R(A_j, A_i),$$

and

$$A_i \sim A_j \Leftrightarrow R(A_i, A_j) = R(A_j, A_i),$$

which imply that the resulting ranking of $A_1, A_2, \cdots, A_n$ can be directly derived from the pairwise comparisons if R is consistent.

Recognizing the insufficiency of one single index in ordering fuzzy numbers, Dubois and Prade proposed four fuzzy relations to indicate the relative locations of two fuzzy numbers [46]. They are:

(1) The grade of possibility of dominance of A_i over A_j

$$PD(A_i, A_j) = \sup_{\substack{x_i, x_j \\ x_i \geq x_j}} \min(A_i(x_i), A_j(x_j));$$

(2) The grade of possibility of strict dominance of A_i over A_j

$$PSD(A_i, A_j) = \sup_{x_i} \inf_{\substack{x_j \\ x_i \leq x_j}} \min(A_i(x_i), 1 - A_j(x_j));$$

(3) The grade of necessity of dominance of A_i over A_j

$$ND(A_i, A_j) = \inf_{x_i} \sup_{\substack{x_j \\ x_i \geq x_j}} \min(1 - A_i(x_i), A_j(x_j));$$

(4) The grade of necessity of strict dominance of A_i over A_j

$$NSD(A_i, A_j) = 1 - \sup_{\substack{x_i, x_j \\ x_i \leq x_j}} \min(A_i(x_i), A_j(x_j)).$$

Clearly, PD is the same as R_{BK} and $NSD(A_i, A_j) = 1 - PD(A_j, A_i)$. Although all the four indices satisfy weak transitivity (Exercise 38) and thus the preceding procedure can be employed to derive a final ranking from each index, these final rankings are not always the same. The final decision is left to the decision maker.

Example 4.12. Let $A = (2, 7, 8)$, $A_2 = (2, 5, 7, 8)$, $A_3 = (2, 5, 8)$ and $A_4 = (2, 3, 8)$. The corresponding indices PD, PSD, ND and NSD are calculated and listed as follows.

PSD	A_1	A_2	A_3	A_4	PD	A_1	A_2	A_3	A_4
A_1	0.5	0.5	0.75	0.83	A_1	1	1	1	1
A_2	0.5	0.5	0.75	0.83	A_2	1	1	1	1
A_3	0.25	0.25	0.5	0.62	A_3	0.75	1	1	1
A_4	0.17	0.17	0.37	0.5	A_4	0.6	0.63	0.63	1

NSD	A_1	A_2	A_3	A_4	ND	A_1	A_2	A_3	A_4
A_1	0	0	0.25	0.4	A_1	0.5	0.63	0.63	0.83
A_2	0	0	0	0.38	A_2	0.37	0.5	0.5	0.75
A_3	0	0	0	0.38	A_3	0.17	0.5	0.5	0.75
A_4	0	0	0	0	A_4	0.17	0.25	0.25	0.5

If we use these indices individually to rank A_1, A_2, A_3, A_4, then the following rankings are obtained.

For PD: $A_1 \sim A_2 \succ A_3 \succ A_4$;
For PSD: $A_1 \sim A_2 \succ A_3 \succ A_4$;
For ND: $A_1 \succ A_2 \sim A_3 \succ A_4$;
For NSD: $A_1 \sim A_2 \succ A_3 \succ A_4$.

Generally, $PD(A, B)$ compares the right spread of A with the left spread of B; $PSD(A, B)$ compares the right spread of A and B; $ND(A, B)$ compares the left spread of A and B; $NSD(A, B)$ compares the left spread of A with the right spread of B. Therefore, the joint use of four indices is somewhat like an illustration of relative position of two fuzzy numbers from multiple angles instead of ranking fuzzy numbers.

4.5.4 Ranking Axioms

A fuzzy number is a set of real numbers with each number corresponding to a degree. A natural idea is to combine these numbers as well as the degrees to form a single number so that comparisons can be based on the order relation of real numbers. The various combination ways have made ranking approaches diverse. Each approach, however, pays attention to a special aspect of a fuzzy number. As a result, each method suffers from some defects if only one number is formed for each fuzzy number. Freeling [60] argues that by reducing the whole of our analysis to a single number, we are losing much of the information we have purposely been keeping throughout our calculations. To avoid the partiality in the use of the single index, Dubois and Prade propose four indices [46] (the so called complete indices) to locate the relative position of two fuzzy numbers in question. As we know, the four indices may derive different final ranking in some cases. The final choice is left to the decision maker. Therefore, Yuan [170] argues that this somehow defeats the purpose of the ranking method which is supposed to derive a conclusion for the decision maker. These opposite standpoints indicate that what is a good method is controversial. In addition, when some concrete fuzzy numbers are given, opposite rankings are frequently obtained by the application of different methods. There are no unified criteria to judge which one is better. In this section, we investigate the reasonability of a ranking method by introducing a set of axioms developed in [149] and examine their fulfillment. For the details, see [149, 150, 151].

Let M be a ranking method and S the set of fuzzy numbers for which the method M can be applied. The following axioms are proposed by combining properties of real numbers and the purpose of ranking fuzzy numbers, which serves decision making analysis.

4.5 Ranking of Fuzzy Numbers

A₁ The relation $\sim$ is reflexive on any finite subset $\mathcal{A}$ of S.

A₂ The relation $\succ$ is asymmetric on any finite subset $\mathcal{A}$ of S.

A₃ The relation $\succsim$ is complete and transitive on any finite subset $\mathcal{A}$ of S.

A₄ For an arbitrary finite subset $\mathcal{A}$ of S and $A, B \in \mathcal{A}$, $\inf \text{supp}(A) > \sup \text{supp}(B)$, we should have $A \succ B$ by M on $\mathcal{A}$.

A₅ Let $\mathcal{A}_1$ and $\mathcal{A}_2$ be two arbitrary finite subsets of S and let A,B be in $\mathcal{A}_1 \cap \mathcal{A}_2$. Then $A \succ B$ by M on $\mathcal{A}_1$ iff $A \succ B$ by M on $\mathcal{A}_2$.

A₆ Let $A, B, A+C$ and $B+C$ be elements in S. If $A \succsim B$ by M on $\{A,B\}$, then $A+C \succsim B+C$ by M on $\{A+C, B+C\}$.

A₇ Let A, B, AC and BC be elements in S and $C \geq 0$. $A \succsim B$ by M on $\{A,B\}$ implies $AC \succsim BC$ by M on $\{AC, BC\}$.

A₁ through **A₃** are self-explanatory.

A₄ means that, if two fuzzy numbers have separate supports, then the fuzzy number with support on the right has a higher ranking than the one with support on the left.

A₅ means that the ranking order of A and B is independent of any other fuzzy number. It is called the independence of irrelevant alternatives (fuzzy numbers in this case) in the framework of decision making analysis.

A₆ indicates that the addition of fuzzy numbers is compatible with $\succsim$ and the explanation is similar for **A₇**.

Clearly, all the other axioms are an extension of properties of real numbers except **A₅**.

In Table 4.1, we list the fulfilment of axioms for some ranking approaches introduced before.

Table 4.1 Fulfilment of axioms for some ranking approaches

index	Y^λ	AD^α	C	K	CH^k	PD	PSD	ND	NSD
A₁	Y	Y	Y	Y	Y	Y	Y	Y	Y
A₂	Y	Y	Y	Y	Y	Y	Y	Y	Y
A₃	Y	Y	Y	Y	Y	Y	Y	Y	Y
A₄	Y	Y	Y	N	Y	Y	Y	Y	Y
A₅	Y	Y	Y	N	N	N	N	N	N
A₆	Y	Y	N	N	N	Y	N	N	Y
A₇	N	Y	N	N	N	Y	N	N	Y

In the following, we give a proof of the fulfilment of axioms for PD. The proof of the others is left to the reader.

Proposition 4.11. *PD satisfies $\mathbf{A_1}$ through $\mathbf{A_7}$ except $\mathbf{A_5}$.*

Proof. The proof that PD satisfies $\mathbf{A_1}$ through $\mathbf{A_3}$ is trivial.

Firstly, we prove that PD satisfies $\mathbf{A_4}$.

If $\inf \operatorname{supp}(A) > \sup \operatorname{supp}(B)$, then $PD(A,B) = 1$ and $PD(B,A) = 0$, and thus $A \succ B$ by Proposition 4.10(4).

Secondly, we prove that PD satisfies $\mathbf{A_6}$.

Assume the kernels of A_i are $[a_i^-, a_i^+]$ ($i = 1, 2, 3$). From $A_1 \succeq A_2$ by PD on $\{A_1, A_2\}$, it follows $a_1^+ \geq a_2^-$, and thus $a_1^+ + a_3^+ \geq a_2^- + a_3^+ \geq a_2^- + a_3^-$. Therefore, $A_1 + A_3 \succeq A_2 + A_3$ by PD on $\{A_1 + A_3, A_2 + A_3\}$.

Finally, we prove that PD satisfies $\mathbf{A_7}$.

Let $\ker(A_i) = [a_i^-, a_i^+]$ ($i = 1, 2, 3$). From $A_1 \succeq A_2$ by PD on $\{A_1, A_2\}$, we have $a_1^+ \geq a_2^-$, and thus $a_1^+ a_3^+ \geq a_2^- a_3^+ \geq a_2^- a_3^-$. Hence, $PD(A_1 A_3, A_2 A_3) \geq PD(A_2 A_3, A_1 A_3)$ and thus $A_1 A_3 \succeq A_2 A_3$ by PD on $\{A_1 A_3, A_2 A_3\}$.

To verify that PD does not satisfy $\mathbf{A_5}$, we present an example in the following.

Consider $A = (0.1, 0.2, 0.3)$, $B = (0.1, 0.2, 0.8, 0.9)$ and $C = (0.7, 0.8, 0.9)$. $\mathcal{A}_1 = \{A, B\}$ and $\mathcal{A}_2 = \{A, B, C\}$. When applying PD on $\mathcal{A}_1$, we get $A \sim B$. When applying PD on $\mathcal{A}_2$, we get $B \succ A$. □

The axiomatic research provides one way to assess the reasonability of a ranking method. However, how to select a complete and proper set of axioms to make the assessment objective is not an easy task. On the one hand, AD^α is a satisfactory index since it fulfils all axioms. On the other hand, this index is far from perfect. For instance, it is heavily dependent on α, it ignores the left spread, it even makes no sense when α is too small, etc. There is a long way to go concerning the reasonability study of ranking methods.

4.6 An Application of Fuzzy Numbers

Let us return to the fuzzy additive weighting formula

$$\tilde{r}_i = \frac{\sum\limits_{j=1}^{m} \tilde{w}_j \tilde{r}_{ij}}{\sum\limits_{j=1}^{m} \tilde{w}_j} \qquad (i = 1, 2, \cdots, n).$$

Assume that $\tilde{r}_{ij} = (r_{ij}^1, r_{ij}^2, r_{ij}^3, r_{ij}^4) \geq 0$ and $\tilde{w}_j = (w_j^1, w_j^2, w_j^3, w_j^4) > 0$ are trapezoidal fuzzy numbers. Then it can be easily verified that

4.6 An Application of Fuzzy Numbers

$$(\tilde{r}_i)_\alpha = \frac{\sum_{j=1}^{m}(\tilde{w}_j)_\alpha(\tilde{r}_{ij})_\alpha}{\sum_{j=1}^{m}(\tilde{w}_j)_\alpha}$$

$$= \left[\frac{A_i + L_{2i}\alpha + L_{1i}\alpha^2}{C + D\alpha}, \frac{D_i + U_{2i}\alpha + U_{1i}\alpha^2}{A + B\alpha}\right],$$

where $A = \sum_{j=1}^{m} w_j^1$, $B = \sum_{j=1}^{m}(w_j^2 - w_j^1)$, $C = \sum_{j=1}^{m} w_j^4$, $D = \sum_{j=1}^{m}(w_j^3 - w_j^4)$,

$A_i = \sum_{j=1}^{m} w_j^1 r_{ij}^1$, $B_i = \sum_{j=1}^{m} w_j r_{ij}^2$, $C_i = \sum_{j=1}^{m} w_j r_{ij}^3$, $D_i = \sum_{j=1}^{m} w_j^4 r_{ij}^4$,

$L_{1i} = \sum_{j=1}^{m}(w_j^2 - w_j^1)(r_{ij}^2 - r_{ij}^1)$, $L_{2i} = \sum_{j=1}^{m}(w_j^1(r_{ij}^2 - r_{ij}^1) + r_{ij}^1(w_j^2 - w_j^1))$,

$U_{1i} = \sum_{j=1}^{m}(w_j^4 - w_j^3)(r_{ij}^4 - r_{ij}^3)$, $U_{2i} = -\sum_{j=1}^{m}(w_j^4(r_{ij}^4 - r_{ij}^3) + r_{ij}^4(w_j^4 - w_j^3))$.

Now we should choose a ranking method to compare the performance evaluations $\tilde{r}_1, \tilde{r}_2, \ldots, \tilde{r}_n$. Considering that the α-cut of $\tilde{r}_n$ is available, we adopt the index Y. In this case,

$$Y(\tilde{r}_i) = \frac{1}{2}\int_0^1 \left(\frac{A_i + L_{2i}\alpha + L_{1i}\alpha^2}{C + D\alpha} + \frac{D_i + U_{2i}\alpha + U_{1i}\alpha^2}{A + B\alpha}\right) d\alpha$$

$$= \frac{L_{1i}}{2D}(\frac{1}{2} + \frac{L_{2i}}{L_{1i}} - \frac{C}{D} + (\frac{A_i}{L_{1i}} + \frac{C^2}{D^2} - \frac{L_{2i}C}{L_{1i}D})\ln(1 + \frac{D}{C})$$

$$+ \frac{U_{1i}}{2B}(\frac{1}{2} + \frac{U_{2i}}{U_{1i}} - \frac{A}{B} + (\frac{D_i}{U_{1i}} + \frac{A^2}{B^2} - \frac{U_{2i}A}{U_{1i}B})\ln(1 + \frac{B}{A}).$$

We will apply the calculating formula to the assessment of radiation protection measures.

Example 4.13. *(Application for the Choice of Radiation Protection Measures)*

In the current literature, some traditional decision making techniques, e.g. cost-benefit analysis, multi-attribute utility analysis and multi-criteria outranking analysis etc. are recommended for the evaluation of available radiation protection measures. The common point of these approaches is the use of crisp numbers. As a matter of fact, some criteria like 'discomfort from ventilation' are merely described in linguistic terms. In order to apply the known decision analysis models, these vague descriptions have to be transformed into numbers without exception. For instance, if one feels 'slightly uncomfortable', the value 0.25 is assigned to the corresponding protection measure. Clearly, the transformation process is the one of information reduction. As a consequence, the results may be distorted. In addition, among other types of uncertainties, the decision maker is confronted with the uncertainties associated with an imperfect knowledge of the performance of the options under different circumstances, or of the parameters and data used in the assessment, e.g. dose estimates, technical specification of the options. The existence of this class of uncertainties also makes the precise descriptions inappropriate. Instead, every evaluation of performance can be regarded as a fuzzy number, and triangular or trapezoidal fuzzy numbers can be used for the description of uncertainties in the evaluations, depending on the degree of the uncertainties included in the evaluating process.

There are five protection options which need to be evaluated according to the following five criteria: annual protection cost, annual collective dose, discomfort from ventilation and annual average individual doses to workers in group. Based on these criteria, all options are evaluated in fuzzy numbers.

Criterion 1: Annual Protection Cost($)

Option 1 (O_1): (10000, 10300, 10550, 11000)
Option 2 (O_2): (17000, 17100, 17400, 18000)
Option 3 (O_3): (18000, 18300, 18600, 19000)
Option 4 (O_4): (31900, 32050, 32350, 32500)
Option 5 (O_5): (35000, 35500, 36000).

Criterion 2: Annual Collective Dose

Option 1: (0.55, 0.561, 0.58)
Option 2: (0.34, 0.357, 0.4)
Option 3: (0.32, 0.335, 0.35)
Option 4: (0.18, 0.196, 0.25)
Option 5: (0.16, 0.178, 0.2).

Criterion 3: Annual Average Individual Doses to Worker in Groups (Group 1)

Option 1: (32.5, 34.5, 35.5)
Option 2: (21, 22.3, 23)
Option 3: (20, 21, 22)
Option 4: (11, 12.6, 14)
Option 5: (10, 11.3, 12).

Criterion 4: Annual Average Individual Doses to Worker in Groups (Group 2)

Option 1: (27, 28.9, 30)
Option 2: (16, 17.1, 18)
Option 3: (15, 16.3, 17.5)
Option 4: (7, 8.4, 9)
Option 5: (7, 7.8, 9).

Criterion 5: Discomfort from Ventilation

For this criterion, the descriptions are in linguistic terms. They are 'no problems' for option 1, 'slightly uncomfortable' for option 2 and option 3, 'severely uncomfortable' for option 4 and 'difficult to work' for option 5.

4.6 An Application of Fuzzy Numbers

These linguistic expressions can be transformed into fuzzy numbers on [0,1], which reflect our satisfaction degrees with options.

Option 1: (1, 1, 1)
Option 2: (0.6, 0.7, 0.8, 0.9)
Option 3: (0.6, 0.7, 0.8, 0.9)
Option 4: (0.1, 0.2, 0.3, 0.4)
Option 5: (0, 0.1, 0.2).

Since the fuzzy numbers for the first three criteria are not in the same scale, we should treat them so that all of them are fuzzy numbers on [0,1]. For the scaling process, we use formula

$$\frac{x_{max} - x}{x_{max} - x_{min}},$$

where x_{max} and x_{min} are the maximal and minimal number respectively for the criterion; and x is the number to be scaled. For example, $x_{max} = 36000$ and $x_{min} = 10000$ for the first criterion. Therefore, the number 10300 is treated as

$$\frac{36000 - 10300}{36000 - 10000} = 0.988462.$$

Note that x_{max} and x_{min} are respectively scaled as 0 and 1. The reason is that a larger number indicates 'less preferable option' for cost and dose criterion. In this way, we obtain all the fuzzy numbers on [0,1] as follows:

Criterion 1

Option 1: (0.962, 0.979, 0.988, 1)
Option 2: (0.692, 0.715, 0.727, 0.731)
Option 3: (0.654, 0.669, 0.681, 0.692)
Option 4: (0.134, 0.14, 0.152, 0.158)
Option 5: (0, 0.019, 0.038).

Criterion 2

Option 1: (0, 0.045, 0.071)
Option 2: (0.429, 0.531, 0.571)
Option 3: (0.548, 0.583, 0.619)
Option 4: (0.786, 0.914, 0.952)
Option 5: (0.922, 0.949, 1).

Criterion 3 (Group 1)

Option 1: (0, 0.039, 0.118)
Option 2: (0.49, 0.518, 0.569)
Option 3: (0.529, 0.568, 0.608)
Option 4: (0.843, 0.898, 0.968)
Option 5: (0.913, 0.965, 1).

Criterion 4 (Group 2)

Option 1: (0, 0.048, 0.13)
Option 2: (0.522, 0.561, 0.609)
Option 3: (0.543, 0.596, 0.652)
Option 4: (0.913, 0.939, 1)
Option 5: (0.913, 0.965, 1).

The fuzzy weights for the criteria are:

$\tilde{w}_1 = (0.15, 0.20, 0.30, 0.35)$
$\tilde{w}_2 = (0.07, 0.08, 0.09, 0.10)$
$\tilde{w}_3 = (0.35, 0.40, 0.45, 0.50)$
$\tilde{w}_4 = (0.04, 0.05, 0.06, 0.07)$
$\tilde{w}_5 = (0.10, 0.13, 0.15, 0.20)$

Applying the preceding formula yields:

$Y(O_1) \approx 0.48, Y(O_2) \approx 0.67, Y(O_3) \approx 0.69, Y(O_4) \approx 0.66, Y(O_5) \approx 0.63.$

Therefore, the ranking of the five options is:

option 3 ≻ option 2 ≻ option 4 ≻ option 5 ≻ option 1,

i.e. option 3 is the best.

4.7 Exercises

1. Prove Proposition 4.2.
2. Let $X = \{a, b, c, d, e, f\}$, $Y = \{u, v, w\}$. A mapping f from X to Y is defined by
$$f(x) = \begin{cases} u & x = a, d, e, f \\ v & x = b, c \end{cases}$$
If $A = 0.1/a + 0.4/b + 1/c + 0.6/d + 0.3/e$, what are $f(A)$, $f(A^c)$, $f(\{a, b\})$, $f(A_{0.6})$ and $(f(A))_{0.6}$.
3. Show that:

   (1) If f is injective, then $\forall A \in F(X), f^{-1}(f(A)) = A$;
   (2) If f is surjective, then $\forall B \in F(Y), f(f^{-1}(B)) = B$.
4. Let $f(x) = \frac{1}{2}x$ and A be defined by
$$A(x) = \begin{cases} x-1 & 1 < x \le 2 \\ 3-x & 2 < x < 3 \\ 0 & \text{otherwise} \end{cases}$$

   Find $f(A)$.

4.7 Exercises

5. Let $f(x) = 1 + \frac{1}{2}(x+1)^2$ and A be defined by
$$A(x) = \begin{cases} 1 + \frac{1}{3}x & -3 < x \leq 0 \\ 1 - x & 0 < x \leq 1 \\ 0 & \text{otherwise} \end{cases}$$
Find $f(A)$.

6. If $A \in F(\mathbb{R})$, apply the extension principle to define the absolute value $|A|$ of A and the opposite $-A$ of A. If $A = 0.3/2 + 0.4/1 + 0.5/0 + 0.1/-1 + 0.8/-2$, find $|A|$ and $-A$.

7. If f is injective, show that: (1) for every $\alpha \in]0,1]$, $(f(A))_\alpha = f(A_\alpha)$ and (2) $\forall A_i \in F(X)$ $(i \in I)$, $f(\bigcap_{i \in I} A_i) = \bigcap_{i \in I} f(A_i)$.

8. Let $A \in F(X)$, $f: X \to Y$ and $g: Y \to Z$. Show that
$$(g \circ f)(A) = \bigcup_{\alpha \in [0,1]} \alpha g(f(A_\alpha)).$$

9. Given $A = 0.5/0 + 1/0.1 + 0.5/0.2$, find $2A$, $A - A$ and $A + A$.

10. Let $f: X_1 \times X_2 \times \cdots \times X_n \to Y$ and $A_i \in F(X_i)(i = 1, 2, \cdots, n)$. Show that $f(A_1, A_2, \cdots, A_n)$ is a normal fuzzy set on Y if $A_1, A_2, \cdots, A_n$ are normal fuzzy sets on $X_1, X_2, \cdots, X_n$ respectively.

11. Show that, for $A_i \in F(X_i)(i = 1, 2, \ldots, n)$, $f: X_1 \times X_2 \times \cdots \times X_n \to Y$,
$$f((A_1)_\alpha, (A_2)_\alpha, \ldots, (A_n)_\alpha) \subseteq (f(A_1, A_2, \ldots, A_n))_\alpha;$$
$$f((A_1)_{\dot\alpha}, (A_2)_{\dot\alpha}, \ldots, (A_n)_{\dot\alpha}) = (f(A_1, A_2, \ldots, A_n))_{\dot\alpha};$$
$$f(A_1, A_2, \ldots, A_n) = \bigcup_{\alpha \in [0,1]} \alpha f((A_1)_\alpha, (A_2)_\alpha, \ldots, (A_n)_\alpha)$$
$$= \bigcup_{\alpha \in [0,1]} \alpha f((A_1)_{\dot\alpha}, (A_2)_{\dot\alpha}, \ldots, (A_n)_{\dot\alpha}).$$

12. If $A_i \in F(X)$ and $B_j \in F(Y)$ $(i \in I, j \in J)$, $f: X \times Y \to Z$, show that
$$f(\bigcup_{i \in I} A_i, \bigcup_{j \in J} B_j) = \bigcup_{i \in I} \bigcup_{j \in J} f(A_i, B_j);$$
$$f(\bigcap_{i \in I} A_i, \bigcap_{j \in J} B_j) \subseteq \bigcap_{i \in I} \bigcap_{j \in J} f(A_i, B_j).$$

13. If $A \in F(\mathbb{R})$ is defined by
$$A(x) = \begin{cases} \frac{(1-x)^2}{4} & -1 \leq x \leq 1 \\ 0 & \text{otherwise} \end{cases},$$
find $A - A$ and $A + A$.

14. Show that $A(B+C) \subseteq AB + AC$ for all $A, B, C \in F(\mathbb{R})$. Give an example to illustrate that the inclusion can not be replaced by the equality generally.
15. Show that $A \in F(\mathbb{R})$ defined by

$$A(x) = \begin{cases} \frac{1}{x} & x \geq 1 \\ 1 & 0 < x < 1 \\ 0 & x \leq 0, \end{cases}$$

is a convex fuzzy quantity.
16. Give an example to illustrate that $A_1 \cup A_2$ is not necessarily a convex fuzzy set when both A_1 and A_2 are convex fuzzy sets.
17. Let A and B be two convex fuzzy quantities. Show that there exists an $x_0 \in \mathbb{R}$ such that

$$(A \vee B)(x) = \begin{cases} A(x) \vee B(x) & x \geq x_0 \\ A(x) \wedge B(x) & x < x_0 \end{cases}.$$

18. Show that if $A, B \in F(\mathbb{R})$, then $\forall x \in \mathbb{R}$,

$$|A(x) - B(x)| = \int_0^1 |A_\alpha(x) - B_\alpha(x)| d\alpha.$$

19. Prove that $A \vee B = A \Leftrightarrow A \wedge B = B$.
20. Given fuzzy numbers $A = (0, 1, 2, 3)$ and $B = (4, 5, 6)$, find $A+B$, $A-B$, AB, A/B.
21. Let w_1, w_2, A_1 and A_2 be non-negative trapezoidal fuzzy numbers. Find a formula to calculate $w_1 A_1 + w_2 A_2$.
22. Let A and B be fuzzy numbers with $\ker(A) = [a^-, a^+]$ and $\ker(B) = [b^-, b^+]$. Define a fuzzy relation R by

$$R(A, B) = \sup_{x \geq y} (A(x) \wedge B(y)).$$

Prove the following statements:
(1) $R(A, B) = \begin{cases} 1 & a^+ \geq b^- \\ \text{hgt}(A \cap B) & a^+ < b^- \end{cases};$
(2) R satisfies weak transitivity and negative transitivity;
(3) R is not transitive.
23. If the Hamming distance between two fuzzy numbers A and B is defined by

$$d_H(A, B) = \int_0^1 |A(x) - B(x)| dx,$$

prove that

$$d_H(A, B) = \int_0^1 d_H(A_\alpha, B_\alpha) d\alpha.$$

4.7 Exercises

24. For a fuzzy number F, define $\bar{F}$ and $\underline{F}$ by

$$\bar{F}(x) = \sup_{y \geq x} F(y) \quad \text{and} \quad \underline{F}(x) = \sup_{y \leq x} F(y).$$

Show that $F = \bar{F} \cap \underline{F}$. If A and B are fuzzy numbers, prove that

$$\underline{A} \vee \underline{B} = \underline{A} \cap \underline{B}, \quad \underline{A} \wedge \underline{B} = \underline{A} \cup \underline{B},$$

$$\bar{A} \wedge \bar{B} = \bar{A} \cap \bar{B}, \quad \bar{A} \vee \bar{B} = \bar{A} \cup \bar{B}.$$

25. Use the results in Exercise 23 and 24 to verify that

$$d_H(\underline{A}, \underline{A} \vee \underline{B}) = \int_0^1 (a_\alpha^- \vee b_\alpha^- - a_\alpha^-) d\alpha,$$

$$d_H(\underline{A}, \underline{A} \wedge \underline{B}) = \int_0^1 (a_\alpha^- - a_\alpha^- \wedge b_\alpha^-) d\alpha,$$

$$d_H(\bar{A}, \bar{A} \vee \bar{B}) = \int_0^1 (a_\alpha^+ \vee b_\alpha^+ - a_\alpha^+) d\alpha,$$

$$d_H(\bar{A}, \bar{A} \wedge \bar{B}) = \int_0^1 (a_\alpha^+ - a_\alpha^+ \wedge b_\alpha^+) d\alpha,$$

where A and B are fuzzy numbers with $A_\alpha = [a_\alpha^-, a_\alpha^+]$ and $B_\alpha = [b_\alpha^-, b_\alpha^+]$.

26. Let $A = (0.5, 0.6, 0.7)$ and $B = (0.1, 0.6, 0.8)$. Use $AD^{0.8}$, Y^λ ($\lambda = 0, 0.5, 1$) and $C(A)$ respectively to rank A and B.

27. Let $A = (a, b, c, d)$ be a trapezoidal fuzzy number. Find the formulas to compute AD^α, Y^λ and $C(A)$ respectively.

28. Chang uses the index $CH(A) = \int_{supp(A)} xA(x)dx$ to evaluate A. A larger $CH(A)$ corresponds to a higher rank. Use this index to rank A and B in Exercise 26.

29. Let A and B be fuzzy numbers. Define the crisp relation R by:

$$ARB \Leftrightarrow A \vee B = A.$$

Prove that R is a partial order. Give an example to illustrate that R is not complete.

30. Let $A = (0.6, 0.7, 0.8)$ and $B = (0.3, 0.7, 1)$. Use the indices K, N_{M_1} and T^1 respectively to rank A and B.

31. Let $A_i = (a_i, b_i, c_i)$ be triangular fuzzy numbers. Find the formulas to compute T^1, $T^{0.5}$ and T^2 respectively. Give an example to illustrate that T^1, $T^{0.5}$ and T^2 may derive three different rankings.

32. Let $A_1 = (3,5,7)$, $A_2 = (4,5,\frac{51}{8})$, $A_3 = (2,3,5)$ and $A_4 = (6,7,8)$. Use T^1 to rank A_1, A_2, A_3, A_4.
33. Let $A_1 = (3,5,7)$, $A_2 = (4,5,\frac{51}{8})$, $A_3 = (2,3,5)$ and $A_4' = (10,11,12)$. Use T^1 to rank A_1, A_2, A_3, A_4'. Compare the result with that of Exercise 32.
34. Show that $R_{BK}(A,B) = N_L(A, A \vee B)$ holds for any fuzzy numbers A and B, where N_L stands for the lattice nearness measure.
35. Show that R^λ is consistent for any $\lambda \in [0,1]$.
36. Show that, for any fuzzy numbers A and B,
    (1) $PD(A,B) + PD(B,A) \geq 1$;
    (2) $PSD(A,B) + PSD(B,A) \leq 1$;
    (3) $ND(A,B) + ND(B,A) = 1$;
    (4) $NSD(A,B) + NSD(B,A) \leq 1$.
37. Show that $(PSD(A,B), PSD(B,A)) \geq 1/2$ for any fuzzy numbers A and B.
38. Show that all the four fuzzy relations PD, PSD, ND and NSD are weakly transitive and none of them is consistent.
39. The application of R^λ and Y^λ derives the same ranking.
40. Let $A = (0.3, 0.5, 1)$ and $B = (0.11, 0.6, 0.8)$. Rank them by using R_{BK}, PD, PSD, ND, NSD respectively.
41. Let $A = (0,1,2,3)$, $B = (0,1,1.5)$ and $C = (1.5,2,3)$. Use R_{BK} to rank A, B and C.
42. Show that if a ranking approach satisfies $\mathbf{A_6}$, then $A \sim B$ implies $A+C \sim B+C$ for any fuzzy number C.
43. Show that if the involved fuzzy numbers are continuous, then both PSD and ND satisfy $\mathbf{A_6}$ and $\mathbf{A_7}$.
44. Prove the remaining results except those related to PD in Table 4.1.
45. Show that $\tilde{r}_i$ in Section 4.6 is a fuzzy number with support $]A_i, D_i[$, kernel $[B_i, C_i]$, left spread
$$f_1(x) = \frac{Dx - L_{2i} + \sqrt{(L_{2i} - Dx)^2 + 4L_{1i}(Cx - A_i)}}{2L_{1i}}$$
and right spread
$$f_2(x) = \frac{Bx - U_{2i} - \sqrt{(U_{2i} - Bx)^2 - 4U_{1i}(D_i - Ax)}}{2U_{1i}},$$
46. In a fuzzy decision tree, the total utility is computed as $p_1 u_1 + p_2 u_2 + \cdots + p_n u_n$ when an alternative is chosen, where p_i is the linguistic probability of occurrence of outcome O_i ($i = 1, 2, \cdots, n$) and u_i is the corresponding fuzzy utility. Suppose that there are two alternatives A_1 and A_2. If A_1 is chosen, then the probability of occurrence of O_1 and O_2 are 'around 0.8' and 'around 0.2' respectively. If A_2 is chosen, then the probability of

occurrence of O_1 and O_2 are 'around 0.4' and 'around 0.6' respectively. When O_1 occurs, the utility is 'approximately between 7 and 8'. When O_2 occurs, the utility is 'approximately between 3 and 4'.

(1) Model the involved linguistic descriptions by using triangular or trapezoidal fuzzy numbers;
(2) Compute the total utility of A_1 and A_2 respectively;
(3) Find a ranking method to determine which one of A_1 and A_2 is better.

Chapter 5
A Brief Introduction to Some Pure Mathematical Topics

As known to us, the theory of classical sets is the foundation on which modern mathematics rests. When sets are fuzzified, some traditional pure mathematical branches are accordingly generalized. In this chapter, we introduce three well-developed fuzzified mathematical areas briefly to have a glance at how a pure mathematical theory can be fuzzified. The three areas are (i) fuzzy measures and fuzzy integrals (ii) fuzzy algebraic structures including fuzzy groups, fuzzy rings and fuzzy fields (iii) fuzzy topology. This chapter will be mainly for authors with the elementary knowledge of the corresponding classical mathematical branches and it will supply them with basic materials for further reading or research.

5.1 Fuzzy Measures and Fuzzy Integrals

There are various definitions of fuzzy measures and integrals in the literature. Sugeno was the first who fuzzified measures and integrals. As a brief introduction, we will start with his definitions [137].

5.1.1 Fuzzy Measures

Definition 5.1. *Let $\mathcal{A} \subseteq P(X)$ be a σ-algebra. If a mapping g from $\mathcal{A}$ to $[0,1]$ satisfies*

(1) boundedness: $g(\emptyset) = 0$ and $g(X) = 1$;
(2) monotonicity: $A \subseteq B$ implies $g(A) \leq g(B)$;
(3) continuity: $A_n \uparrow$ (or $\downarrow$)A implies that $\lim_{n \to \infty} g(A_n) = g(A)$,

then g is called a fuzzy measure. Meanwhile, $(X, \mathcal{A})$ and $(X, \mathcal{A}, g)$ are called a fuzzy measurable space and a fuzzy measure space respectively.

Sugeno [137] made the following interpretation: $g(A)$ measures the certainty degree to which a generic element x is in A. If A is empty, x is certainly not in A. If A is the whole set, x is certainly in it. When $A \subseteq B$, the certainty degree to which x is in A is of course less than the certainty degree to which x is in B.

Example 5.1. *Each probability measure P is a fuzzy measure.*

Since a probability measure apparently satisfies the requirements (1) and (2) in Definition 5.1, it suffices to prove (3). Let $\{A_n\}$ be a monotone increasing sequence, that is, $A_1 \subseteq A_2 \subseteq \cdots$. Write $A = \bigcup_{i=1}^{\infty} A_i$. Then $A = A_1 \cup (A_2 - A_1) \cup (A_3 - A_2) \cup \cdots$. As a probability measure P is countably additive and subtractive, we get successively:

$$P(A) = P(A_1) + P(A_2 - A_1) + P(A_3 - A_2) + \cdots$$
$$= P(A_1) + P(A_2) - P(A_1) + P(A_3) - P(A_2) + \cdots$$
$$= \lim_{n \to \infty} P(A_n).$$

The proof of continuity is similar when $\{A_n\}$ is a monotone decreasing sequence.

Example 5.2. *For any A in $P(X)$, the Dirac measure centered in $x_0 \in X$ assumes the form*

$$g(A) = A(x_0) = \begin{cases} 1 & x_0 \in A \\ 0 & x_0 \notin A \end{cases},$$

where x_0 is a fixed element in X. It can be verified that the Dirac measure is a fuzzy measure.

Example 5.3. *For $h \in F(X)$ and $A \in P(X)$, the possibility measure of A is defined by*

$$g(A) = \bigvee_{x \in A} h(x).$$

If h is normal, then g is a fuzzy measure.

Proposition 5.1. *If g is a fuzzy measure on the measurable space $(X, \mathcal{A})$ and $A, B \in \mathcal{A}$, then*

(1) $g(A \cup B) \geq g(A) \vee g(B)$;
(2) $g(A \cap B) \leq g(A) \wedge g(B)$.

Proof. They are direct consequences of Definition 5.1(2). □

More generally,

$$g(\bigcup_{n=1}^{\infty} A_n) \geq \bigvee_{n=1}^{\infty} g(A_n)$$

$$g(\bigcap_{n=1}^{\infty} A_n) \leq \bigwedge_{n=1}^{\infty} g(A_n).$$

5.1 Fuzzy Measures and Fuzzy Integrals

Definition 5.2. *If a mapping $g_\lambda : \mathcal{A} \to [0,1]$ depending on a parameter λ ($\lambda > -1$) satisfies that*

(1) $g_\lambda(X) = 1$,
(2) $g_\lambda(A \cup B) = g_\lambda(A) + g_\lambda(B) + \lambda g_\lambda(A) g_\lambda(B)$ *whenever* $A \cap B = \emptyset$,
(3) $A_n \uparrow (\downarrow) A$ *implies that* $\lim_{n \to \infty} g_\lambda(A_n) = g_\lambda(A)$,

then g_λ is called a λ-fuzzy measure or a g_λ measure.

Proposition 5.2. *Each g_λ measure is a fuzzy measure.*

Proof. It follows from $X \cap \emptyset = \emptyset$ and $X \cup \emptyset = X$ that

$$g_\lambda(X) = g_\lambda(X \cup \emptyset) = g_\lambda(X) + g_\lambda(\emptyset) + \lambda g_\lambda(X) g_\lambda(\emptyset),$$

i.e. $(1+\lambda) g_\lambda(\emptyset) = 0$ or since $\lambda > -1$, $g_\lambda(\emptyset) = 0$.

Assume that $A \subseteq B$. Then $A \cup (B - A) = A \cup B = B$, together with $A \cap (B - A) = \emptyset$ leads to

$$g_\lambda(B) = g_\lambda(A) + g_\lambda(B - A) + \lambda g_\lambda(A) g_\lambda(B - A) \geq g_\lambda(A).$$

Taking (1) and (3) into account, we obtain that a g_λ measure is a fuzzy measure. □

Proposition 5.3. *Each g_λ measure satisfies the following properties.*

(1) $g_\lambda(A^c) = \dfrac{1 - g_\lambda(A)}{1 + \lambda g_\lambda(A)}$.

(2) *If* $A \supseteq B$, *then* $g_\lambda(A - B) = \dfrac{g_\lambda(A) - g_\lambda(B)}{1 + \lambda g_\lambda(B)}$.

(3) *If* $A_i \cap A_j = \emptyset$ ($i \neq j$), *then*

$$g_\lambda\left(\bigcup_{n=1}^{\infty} A_n\right) = \frac{1}{\lambda}\left(\prod_{n=1}^{\infty}(1 + \lambda g_\lambda(A_n)) - 1\right).$$

Proof. (1) From $A \cap A^c = \emptyset$, we have $g_\lambda(A \cup A^c) = g_\lambda(A) + g_\lambda(A^c) + \lambda g_\lambda(A) g_\lambda(A^c)$. The desired equality follows from $g_\lambda(A \cup A^c) = g_\lambda(X) = 1$.
(2) Suppose $A \supseteq B$,

$$g_\lambda(A) = g_\lambda(B \cup (A - B)) = g_\lambda(B) + g_\lambda(A - B) + \lambda g_\lambda(B) g_\lambda(A - B),$$

i.e.

$$g_\lambda(A - B) = \frac{g_\lambda(A) - g_\lambda(B)}{1 + \lambda g_\lambda(B)}.$$

(3) From $A_1 \cap A_2 = \emptyset$, it follows that

$$g_\lambda(A_1 \cup A_2) = g_\lambda(A_1) + g_\lambda(A_2) + \lambda g_\lambda(A_1) g_\lambda(A_2)$$
$$= \frac{1}{\lambda}((1 + \lambda g_\lambda(A_1))(1 + \lambda g_\lambda(A_2)) - 1).$$

By mathematical induction, we can derive the general formula

$$g_\lambda(\bigcup_{i=1}^{n} A_i) = \frac{1}{\lambda}(\prod_{i=1}^{n}(1 + \lambda g_\lambda(A_i)) - 1).$$

As $n \to \infty$, we obtain

$$g_\lambda(\bigcup_{i=1}^{\infty} A_i) = \frac{1}{\lambda}(\prod_{i=1}^{\infty}(1 + \lambda g_\lambda(A_i)) - 1),$$

due to the continuity of g_λ. □

Proposition 5.4. *For $A, B \in \mathcal{A}$,*

$$g_\lambda(A \cup B) = \frac{g_\lambda(A) + g_\lambda(B) - g_\lambda(A \cap B) + \lambda g_\lambda(A) g_\lambda(B)}{1 + \lambda g_\lambda(A \cap B)}$$

Proof. On the one hand,

$$g_\lambda(A \cup B) = g_\lambda(A \cup (B - A)) = g_\lambda(A) + g_\lambda(B - A) + \lambda g_\lambda(A) g_\lambda(B - A) \quad (5.1)$$

On the other hand, $B = (B \cap A) \cup (B \cap A^c)$ and $(B \cap A) \cap (B \cap A^c) = \emptyset$. Hence,

$$g_\lambda(B) = g_\lambda(A \cap B) + g_\lambda(B - A) + \lambda g_\lambda(A \cap B) g_\lambda(B - A) \quad (5.2)$$

By eliminating $g_\lambda(B - A)$ in (5.1) and (5.2), the desired equality is obtained. □

Example 5.4. *Let $X = \{x_1, x_2, \ldots, x_n\}$ and $\mathcal{A} = P(X)$. If $g_i \in [0, 1]$ $(i = 1, 2, \ldots, n)$ satisfies that*

$$\prod_{i=1}^{n}(1 + \lambda g_i) = 1 + \lambda \quad (5.3)$$

then g_λ defined by $\forall A \in \mathcal{A}$,

$$g_\lambda(A) = \frac{1}{\lambda}(\prod_{x_i \in A}(1 + \lambda g_i) - 1) \quad (5.4)$$

is a λ-fuzzy measure.

Conversely, if g_λ is a λ-fuzzy measure, then (5.3) and (5.4) hold for $g_i = g_\lambda(\{x_i\})$ $(i = 1, 2, \ldots, n)$.

Proof. Assume that (5.3) and (5.4) are satisfied. Then

$$g_\lambda(X) = \frac{1}{\lambda}(\prod_{i}^{n}(1 + \lambda g_i) - 1) = \frac{1}{\lambda}(1 + \lambda - 1) = 1.$$

5.1 Fuzzy Measures and Fuzzy Integrals

Suppose that $A \cap B = \emptyset$. Write $a = \prod_{x_i \in A}(1 + \lambda g_i)$ and $b = \prod_{x_i \in B}(1 + \lambda g_i)$. From (5.4), we know that

$$g_\lambda(A \cup B) = \frac{1}{\lambda}\left(\prod_{x_i \in A \cup B}(1 + \lambda g_i) - 1\right) = \frac{1}{\lambda}\left(\prod_{x_i \in A}(1 + \lambda g_i)\prod_{x_i \in B}(1 + \lambda g_i) - 1\right)$$

$$= \frac{1}{\lambda}(ab - 1) = \frac{1}{\lambda}(a - 1) + \frac{1}{\lambda}(b - 1) + \lambda\frac{1}{\lambda}(a - 1)\frac{1}{\lambda}(b - 1)$$

$$= g_\lambda(A) + g_\lambda(B) + \lambda g_\lambda(A)g_\lambda(B).$$

Since X is a finite set, the continuity requirement is automatically satisfied. Thus g_λ is a λ-fuzzy measure.

Conversely, assume that g_λ is a λ-fuzzy measure.

$$g_\lambda(\{x_1, x_2\}) = g_\lambda(\{x_1\}) + g_\lambda(\{x_2\}) + \lambda g_\lambda(\{x_1\})g_\lambda(\{x_2\})$$

$$= g_1 + g_2 + \lambda g_1 g_2 = \frac{1}{\lambda}((1 + \lambda g_1)(1 + \lambda g_2) - 1).$$

Hence (5.4) is true for $A = \{x_1, x_2\}$. Applying mathematical induction, we can prove (5.4). Observe that $g_\lambda(X) = 1$. By (5.4), $\frac{1}{\lambda}(\prod_{i}^{n}(1 + \lambda g_i) - 1) = 1$, which is the equality (5.3).

5.1.2 Fuzzy Integrals

Definition 5.3. *Let $(X, \mathcal{A}, g)$ be a fuzzy measure space and let $h : X \to [0, 1]$ be a measurable function on $(X, \mathcal{A})$. The fuzzy integral of h over A ($A \in \mathcal{A}$), denoted by $\int_A h dg$, is defined as*

$$\int_A h dg = \bigvee_{\alpha \in [0,1]} (\alpha \wedge g(h_\alpha \cap A)),$$

where h_α stands for the α-cut of h.

Theorem 5.1

$$\int_A h dg = \bigvee_{E \in \mathcal{A}} ((\bigwedge_{x \in E} h(x)) \wedge g(E \cap A)).$$

Proof. On the one hand, by putting $\lambda_E = \bigwedge_{x \in E} h(x)$, we have $E \subseteq h_{\lambda_E}$. Thus

$$\bigvee_{E \in \mathcal{A}} ((\bigwedge_{x \in E} h(x)) \wedge g(E \cap A)) \leq \bigvee_{E \in \mathcal{A}} (\lambda_E \wedge g(h_{\lambda_E} \cap A))$$

$$\leq \bigvee_{\alpha \in [0,1]} (\alpha \wedge g(h_\alpha \cap A)),$$

i.e.
$$\bigvee_{E\in\mathcal{A}}(\bigwedge_{x\in E} h(x)) \wedge g(E\cap A)) \leq \int_A h dg.$$

On the other hand, since $\bigwedge_{x\in h_\alpha} h(x) \geq \alpha$ and $h_\alpha \in \mathcal{A}$ ($\alpha \in [0,1]$),

$$\int_A h dg = \bigvee_{\alpha\in[0,1]}(\alpha \wedge g(h_\alpha \cap A))$$
$$\leq \bigvee_{\alpha\in[0,1]}((\bigwedge_{x\in h_\alpha} h(x)) \wedge g(h_\alpha \cap A))$$
$$\leq \bigvee_{E\in\mathcal{A}}((\bigwedge_{x\in E} h(x)) \wedge g(E\cap A)).$$

Therefore,
$$\int_A h dg = \bigvee_{E\in\mathcal{A}}(\bigwedge_{x\in E} h(x)) \wedge g(E\cap A)). \qquad \square$$

Theorem 5.2
$$\int_A h dg = \bigvee_{\alpha\in[0,1]}(\alpha \wedge g(h_{\dot\alpha} \cap A)).$$

Proof. Clearly,
$$\int_A h dg = \bigvee_{\alpha\in[0,1]}(\alpha \wedge g(h_\alpha \cap A)) \geq \bigvee_{\alpha\in[0,1]}(\alpha \wedge g(h_{\dot\alpha} \cap A))$$

due to the monotonicity of g. Suppose
$$\int_A h dg > \bigvee_{\alpha\in[0,1]}(\alpha \wedge g(h_{\dot\alpha} \cap A)).$$

Then a number $\beta \in [0,1]$ can be found such that
$$\int_A h dg > \beta > \bigvee_{\alpha\in[0,1]}(\alpha \wedge g(h_{\dot\alpha} \cap A)) \geq \alpha \wedge g(h_{\dot\alpha} \cap A)$$

for any α in $[0,1]$.

There must be $g(h_{\dot\beta} \cap A) < \beta$. Since $h_\alpha \subseteq h_{\dot\beta}$ for $\alpha > \beta$,

$$\bigvee_{\alpha\in[0,1]}(\alpha \wedge g(h_\alpha \cap A))$$
$$= (\bigvee_{\alpha\leq\beta}(\alpha \wedge g(h_\alpha \cap A))) \vee (\bigvee_{\alpha>\beta}(\alpha \wedge g(h_\alpha \cap A)))$$
$$\leq \beta \vee \bigvee_{\alpha>\beta}(\alpha \wedge g(h_{\dot\beta} \cap A)) = \beta,$$

which is in contradiction with $\int_A h dg > \beta$. $\qquad \square$

5.1 Fuzzy Measures and Fuzzy Integrals

Definition 5.4. *Let $\{A_i\}$ be a mutually disjoint finite sequence of sets in $\mathcal{A}$ such that $\bigcup_{i=1}^{n} A_i = X$. If a mapping s from X to $[0,1]$ satisfies that $\forall x \in X$, $s(x) = \bigvee_{i=1}^{n} (\alpha_i \wedge A_i(x))$ ($\alpha_i \in [0,1]$), then s is called a simple function. The set of simple functions is denoted by φ.*

For every $A \in \mathcal{A}$ and $s(x) = \bigvee_{i=1}^{n}(\alpha_i \wedge A_i(x))$, write $H_A(s) = \bigvee_{i=1}^{n}(\alpha_i \wedge g(A \cap A_i))$.

Theorem 5.3. *For any $h \in F(X)$ and $E \in \mathcal{A}$,*

$$\int_E h\, dg = \bigvee_{\substack{s \in \varphi \\ s \subseteq h}} H_E(s).$$

Proof. By Theorem 5.1, it suffices to prove

$$\bigvee_{A \in \mathcal{A}} (\bigwedge_{x \in A} h(x)) \wedge g(E \cap A)) = \bigvee_{\substack{s \in \varphi \\ s \subseteq h}} H_E(s).$$

For every $A \in \mathcal{A}$, let $\lambda_A = \bigwedge_{x \in A} h(x)$.
Then $s_0(x) = \lambda_A \wedge A(x)$ is a simple function such that $s_0 \subseteq h$. Thus $\forall A \in \mathcal{A}$,

$$\bigvee_{\substack{s \in \varphi \\ s \subseteq h}} H_E(s) \geq H_E(s_0) = \lambda_A \wedge g(E \cap A) = (\bigwedge_{x \in A} h(x)) \wedge g(E \cap A)).$$

As a consequence,

$$\bigvee_{\substack{s \in \varphi \\ s \subseteq h}} H_E(s) \geq \bigvee_{A \in \mathcal{A}} (\bigwedge_{x \in A} h(x)) \wedge g(E \cap A)).$$

On the other hand, let $s(x) = \bigvee_{i=1}^{n}(\alpha_i \wedge A_i(x))$ be a simple function contained in h. Then there exists an i_0 ($1 \leq i_0 \leq n$) such that

$$H_E(s) = \bigvee_{i=1}^{n}(\alpha_i \wedge g(E \cap A_i)) = \alpha_{i_0} \wedge g(A_{i_0} \cap E).$$

Since

$$\alpha_{i_0} \wedge A_{i_0}(x) \leq \bigvee_{i=1}^{n}(\alpha_i \wedge A_i(x)) \leq h(x),$$

we have $\alpha_{i_0} \leq \bigwedge_{x \in A_{i_0}} h(x)$. Consequently,

$$H_E(s) \leq (\bigwedge_{x \in A_{i_0}} h(x)) \wedge g(A_{i_0} \cap E) \leq \bigvee_{A \in \mathcal{A}} (\bigwedge_{x \in A} h(x)) \wedge g(E \cap A)).$$

That is,
$$\bigvee_{\substack{s \in \varphi \\ s \subseteq h}} H_E(s) \leq \bigvee_{A \in \mathcal{A}} (\bigwedge_{x \in A} h(x)) \wedge g(E \cap A)). \qquad \square$$

In the following, assume that $(X, \mathcal{A}, g)$ is a fuzzy measure space and h is a measurable function on $(X, \mathcal{A})$. We investigate the properties of fuzzy integrals.

Property 5.1
$$0 \leq \int_A h dg \leq 1.$$

Proof. Trivial.

Property 5.2
$$\int_A c\, dg = c \wedge g(A) \text{ with } c \in [0,1].$$

Proof. By Theorem 5.1,
$$\int_A c\, dg = \bigvee_{E \in \mathcal{A}}(\bigwedge_{x \in E} c) \wedge g(E \cap A)) = \bigvee_{E \in \mathcal{A}}(c \wedge g(E \cap A)) = c \wedge g(A). \qquad \square$$

Property 5.3. *(1) If $A \subseteq B$ ($A, B \in \mathcal{A}$), then*
$$\int_A h dg \leq \int_B h dg.$$

(2) If h_1 and h_2 are measurable functions on $(X, \mathcal{A})$ and $h_1 \subseteq h_2$, then
$$\int_A h_1 dg \leq \int_A h_2 dg.$$

Proof. They follow immediately from the definition of fuzzy integrals and the monotonicity of a fuzzy measure. $\qquad \square$

Property 5.4. *For $A, B \in \mathcal{A}$,*
$$\int_{A \cup B} h dg \geq (\int_A h dg) \vee (\int_B h dg).$$
$$\int_{A \cap B} h dg \leq (\int_A h dg) \wedge (\int_B h dg).$$

Proof. They are direct results of Property 5.3(1). $\qquad \square$

Property 5.5. *For any measurable functions h_1 and h_2 on $(X, \mathcal{A})$,*
$$\int_A (h_1 \vee h_2) dg \geq (\int_A h_1 dg) \vee (\int_A h_2 dg).$$

5.1 Fuzzy Measures and Fuzzy Integrals

$$\int_A (h_1 \wedge h_2) dg \leq (\int_A h_1 dg) \wedge (\int_A h_2 dg).$$

Proof. They are direct results of Property 5.3(2). □

Property 5.6. *For any $\alpha \in [0,1]$,*

$$\int_A (\alpha \vee h) dg = (\alpha \wedge g(A)) \vee \int_A h dg.$$

$$\int_A (\alpha \wedge h) dg = \alpha \wedge \int_A h dg.$$

Proof

$$\int_A (\alpha \vee h) dg$$
$$= \bigvee_{E \in \mathcal{A}} (\bigwedge_{x \in E} (\alpha \vee h(x)) \wedge g(E \cap A)) \quad \text{(by Theorem 5.1)}$$
$$= \bigvee_{E \in \mathcal{A}} ((\alpha \vee (\bigwedge_{x \in E} h(x))) \wedge g(E \cap A))$$
$$= (\bigvee_{E \in \mathcal{A}} (\alpha \wedge g(E \cap A))) \vee (\bigvee_{E \in \mathcal{A}} ((\bigwedge_{x \in E} h(x)) \wedge g(E \cap A)))$$
$$= (\alpha \wedge g(A)) \vee \int_A h dg \quad \text{(by Theorem 5.1)}.$$

Similarly,

$$\int_A (\alpha \wedge h) dg = \bigvee_{E \in \mathcal{A}} (\bigwedge_{x \in E} (\alpha \wedge h(x))) \wedge g(E \cap A))$$
$$= \bigvee_{E \in \mathcal{A}} (\alpha \wedge (\bigwedge_{x \in E} h(x))) \wedge g(E \cap A))$$
$$= \alpha \wedge \bigvee_{E \in \mathcal{A}} ((\bigwedge_{x \in E} h(x)) \wedge g(E \cap A)))$$
$$= \alpha \wedge \int_A h dg.$$
□

Property 5.7. *If $\int_E h dg = 0$, then $g(E \cap supp(h)) = 0$.*

Proof. Write
$$A_n = \{x | h(x) \geq \frac{1}{n}\}.$$

Then $A_n \uparrow A = \{x | h(x) > 0\}$. Hence

$$0 = \int_E h dg \geq \int_{E \cap A_n} h dg \geq \int_{E \cap A_n} \frac{1}{n} dg = \frac{1}{n} \wedge g(E \cap A_n).$$

Consequently, $g(E \cap A_n) = 0$. Therefore,

$$g(E \cap supp(h)) = \lim_{n \to \infty} g(E \cap A_n) = 0. \qquad \square$$

Finally, we prove two convergence theorems that generalize their classical counterparts.

Theorem 5.4. *If $\{h_n(x)\}$ is a monotone sequence of measurable functions on $(X, \mathcal{A})$ for every $x \in X$ and h is a measurable function on $(X, \mathcal{A})$, then $\lim_{n \to \infty} h_n(x) = h(x)$ implies that $\lim_{n \to \infty} \int_A h_n dg = \int_A h dg$ for any A in $\mathcal{A}$.*

Proof. We prove the theorem under the assumption that $\{h_n(x)\}$ is a monotone increasing sequence for every x.

By this assumption and Property 5.3(2), $\{\int_A h_n dg\}$ is an increasing sequence. Hence

$$\begin{aligned}
\lim_{n \to \infty} \int_A h_n dg &= \bigvee_{n=1}^{\infty} \int_A h_n dg \\
&= \bigvee_{n=1}^{\infty} \bigvee_{\alpha \in [0,1]} (\alpha \wedge g(A \cap (h_n)_{\underset{\sim}{\alpha}})) \\
&= \bigvee_{\alpha \in [0,1]} (\alpha \wedge \bigvee_{n=1}^{\infty} g(A \cap (h_n)_{\underset{\sim}{\alpha}})) \\
&= \bigvee_{\alpha \in [0,1]} (\alpha \wedge \lim_{n \to \infty} g(A \cap (h_n)_{\underset{\sim}{\alpha}})) \\
&= \bigvee_{\alpha \in [0,1]} (\alpha \wedge g(\lim_{n \to \infty} A \cap (h_n)_{\underset{\sim}{\alpha}})) \\
&= \bigvee_{\alpha \in [0,1]} (\alpha \wedge g(A \cap \bigcup_{n=1}^{\infty} (h_n)_{\underset{\sim}{\alpha}}))
\end{aligned}$$

It is easy to prove that $\bigcup_{n=1}^{\infty} (h_n)_{\underset{\sim}{\alpha}} = h_{\underset{\sim}{\alpha}}$.

Therefore

$$\lim_{n \to \infty} \int_A h_n dg = \bigvee_{\alpha \in [0,1]} (\alpha \wedge (g(A \cap h_{\underset{\sim}{\alpha}})) = \int_A h dg.$$

Theorem 5.5

$$\int_A \underline{\lim}_{n \to \infty} h_n dg \leq \underline{\lim}_{n \to \infty} \int_A h_n dg.$$

Proof

$$\begin{aligned}
\int_A \underline{\lim}_{n \to \infty} h_n dg &= \int_A \lim_{n \to \infty} \inf_{k \geq n} h_k dg \\
&= \lim_{n \to \infty} \int_A \inf_{k \geq n} h_k(x) dg \quad \text{(by Theorem 5.4)} \\
&\leq \lim_{n \to \infty} \inf_{k \geq n} \int_A h_k dg \\
&= \underline{\lim}_{n \to \infty} \int_A h_n dg. \qquad \square
\end{aligned}$$

Remark 5.1. *The definition of fuzzy integrals was firstly proposed in his Ph. D. thesis by Sugeno [137] in 1974 and was extended by Ralescu and Adams [121] in 1986. Some modified versions such as (N)-fuzzy integral and (H)-fuzzy integral can be found in [66, 181]. In 1986, Suarez and Gilavarez [136] gave the concept of a generalized fuzzy integral by using t-norms. In 1990, Wu et al. [161] presented the G-fuzzy integral which contains the Sugeno integral and (N)-fuzzy integral as special cases. Concerning the relationships between different convergence concepts of sequences of fuzzy measurable functions and various convergence theorems of fuzzy integrals, see [153, 154, 155, 156]. For the fuzzy measure and fuzzy integral on L-fuzzy sets, the reader may refer to [178].*

5.2 Fuzzy Algebra

In 1971, Rosenfeld [118] published the seminal paper entitled *Fuzzy Groups*. From then on, various algebraic structures have been fuzzified and thus extended. In this section, we merely introduce the fuzzification of some main notions in abstract algebra including groups, normal groups, rings and ideals.

5.2.1 Fuzzy Subgroups

In this subsection, G denotes an arbitrary group with a multiplicative binary operation and identity e. We first introduce the notion of a fuzzy subgroup.

Definition 5.5. *A fuzzy subset A on G is called a fuzzy subgroup of G if it satisfies the following conditions:*

(1) $A(xy) \geq A(x) \wedge A(y)$ for any $x, y \in G$ and
(2) $A(x^{-1}) \geq A(x)$ for any $x \in G$.

As we know, a subset A of group G is a subgroup of G iff G satisfies that (1) $x, y \in A$ implies $xy \in A$ and (2) $x \in A$ implies $x^{-1} \in A$. The two inequalities in Definition 5.5 are just the fuzzification of these conditions.

Let A be a fuzzy subgroup of G. It follows immediately from the definition that, for any $x \in G$,

(1) $A(x) \leq A(e)$,
(2) $A(x^{-1}) = A(x)$,
(3) $A(x^n) \geq A(x)$, where n is an arbitrary integer.

Proposition 5.5. *Let $A \in F(G)$. Then A is a fuzzy subgroup of G iff $A(xy^{-1}) \geq A(x) \wedge A(y)$ holds for any $x, y \in G$.*

Proof. If A is a fuzzy subgroup of G, then

$$A(xy^{-1}) \geq A(x) \wedge A(y^{-1}) = A(x) \wedge A(y).$$

Conversely, suppose $A(xy^{-1}) \geq A(x) \wedge A(y)$ holds for any $x, y \in G$. Then for any $x \in G$, $A(e) = A(xx^{-1}) \geq A(x) \wedge A(x) = A(x)$, i.e. $A(x) \leq A(e)$. Thus, for any $x \in G$,

$$A(x^{-1}) = A(ex^{-1}) \geq A(e) \wedge A(x) = A(x).$$

Meanwhile, for any $x, y \in G$,

$$A(xy) = A(x(y^{-1})^{-1}) \geq A(x) \wedge A(y^{-1}) \geq A(x) \wedge A(y).$$

Therefore A is a fuzzy subgroup of G. □

In the following proposition we prove the characterization of a fuzzy subgroup in terms of its α-cuts.

Proposition 5.6. *A is a fuzzy subgroup of G iff A_α is a subgroup of G for every $\alpha \in rng(G)$.*

Proof. Suppose that A is a fuzzy subgroup of G and $\alpha \in rng(G)$. Then $A_\alpha \neq \emptyset$. Let $x, y \in A_\alpha$, i.e. $A(x) \geq \alpha$ and $A(y) \geq \alpha$. Hence,

$$A(xy^{-1}) \geq A(x) \wedge A(y^{-1}) = A(x) \wedge A(y) \geq \alpha,$$

and thus $xy^{-1} \in A_\alpha$. As a result, A_α is a subgroup of G.

Conversely, suppose A_α is a subgroup of G for every $\alpha \in A(G)$. For any $x, y \in G$, let $\alpha = A(x) \wedge A(y) \in A(G)$. Then $A(x) \geq \alpha$ and $A(y) \geq \alpha$, i.e. $x \in A_\alpha$ and $y \in A_\alpha$. Hence, $xy^{-1} \in A_\alpha$ since A_α is a subgroup of G. Consequently,

$$A(xy^{-1}) \geq \alpha = A(x) \wedge A(y).$$

By Proposition 5.5, A is a fuzzy subgroup of G. □

Particularly, $A_{A(e)} = \{x | A(x) = A(e)\}$ is a subgroup of G if A is a fuzzy subgroup of G. We shall denote this subgroup by A^*.

The binary multiplicative operation in G can be extended to $F(G)$ using the Zadeh's extension principle. Let $A, B \in F(G)$. Then $A \circ B$ is defined by: for any $z \in G$,

$$(A \circ B)(z) = \bigvee_{z=xy} (A(x) \wedge B(y)).$$

In addition, for every $A \in F(G)$, we shall define $A^{-1} \in F(G)$ by: for any $x \in G$, $A^{-1}(x) = A(x^{-1})$. With these notions, we present an equivalent statement of a fuzzy subgroup.

Proposition 5.7. *Let $A \in F(G)$. Then A is a fuzzy subgroup of G iff $A \circ A^{-1} = A$.*

5.2 Fuzzy Algebra

Proof. If A is a fuzzy subgroup of G, then for any $z \in G$,

$$(A \circ A^{-1})(z) = \bigvee_{z=xy} (A(x) \wedge A^{-1}(y)) = \bigvee_{z=xy} (A(x) \wedge A(y))$$
$$\leq \bigvee_{z=xy} A(xy) = A(z).$$

Hence $A \circ A^{-1} \subseteq A$.

Meanwhile, for any $z \in G$,

$$(A \circ A^{-1})(z) = \bigvee_{z=xy} (A(x) \wedge A^{-1}(y)) = \bigvee_{z=xy} (A(x) \wedge A(y))$$
$$\geq A(z) \wedge A(e) = A(z).$$

Thus $A \circ A^{-1} \supseteq A$. Consequently, $A \circ A^{-1} = A$.

Conversely, suppose $A \circ A^{-1} = A$. Then, for any $x, y \in G$,

$$A(xy^{-1}) = (A \circ A^{-1})(xy^{-1}) \geq A(x) \wedge A^{-1}(y^{-1}) = A(x) \wedge A(y).$$

By Proposition 5.5, A is a fuzzy subgroup of A. $\square$

Proposition 5.8. *Let A be a fuzzy subgroup of G and let f be an epimorphism of G onto a group G'. Then $f(A)$ is a fuzzy subgroup of G'.*

Proof. Let $u, v \in G'$. Since f is surjective, there exist some $x, y \in G$ such that $f(x) = u$ and $f(y) = v$. Hence, we successively obtain:

$$f(A)(u) \wedge f(A)(v) = \left(\bigvee_{f(x)=u} A(x) \right) \wedge \left(\bigvee_{f(y)=v} A(y) \right)$$
$$= \bigvee_{f(x)=u, f(y)=v} (A(x) \wedge A(y))$$
$$\leq \bigvee_{f(x)=u, f(y)=v} A(xy) \quad (A \text{ is a fuzzy subgroup})$$
$$\leq \bigvee_{f(x)f(y)=uv} A(xy)$$
$$= \bigvee_{f(xy)=uv} A(xy) \quad (f \text{ is a homomorphism})$$
$$= \bigvee_{f(z)=uv} A(z) = f(A)(uv).$$

In addition, for any $u \in G'$,

$$f(A)(u^{-1}) = \bigvee_{f(z)=u^{-1}} A(z) = \bigvee_{f(z^{-1})=u} A(z) \quad (f \text{ is a homomorphism})$$
$$= \bigvee_{f(z)=u} A(z^{-1}) = \bigvee_{f(z)=u} A(z) \quad (A \text{ is a fuzzy subgroup})$$
$$= f(A)(u).$$

Hence $f(A)$ is a fuzzy subgroup of G'. $\square$

Proposition 5.9. *Let f be a homomorphism from G to a group G' and let B be a fuzzy subgroup of G'. Then $f^{-1}(B)$ is a fuzzy subgroup of G.*

Proof. For any $x, y \in G$, $f^{-1}(B)(xy) = B(f(xy)) = B(f(x)f(y)) \geq B(f(x)) \wedge B(f(y)) = f^{-1}(B)(x) \wedge f^{-1}(B)(y)$. Meanwhile, for any $x \in G$, $f^{-1}(B)(x^{-1}) = B(f(x^{-1})) = B((f(x))^{-1}) = B(f(x)) = f^{-1}(B)(x)$. Hence $f^{-1}(B)$ is a fuzzy subgroup of G. □

Finally, we briefly introduce the product of fuzzy subgroups. Let $G_1, G_2, \cdots, G_n$ be n groups. We know from abstract algebra that $G_1 \times G_2 \times \cdots \times G_n$ is still a group under the multiplication defined by: $\forall x_i, y_i \in G_i (i = 1, 2, \cdots, n)$

$$(x_1, x_2, \cdots, x_n)(y_1, y_2, \cdots, y_n) = (x_1 y_1, x_2 y_2, \cdots, x_n y_n).$$

In this group, $(x_1, x_2, \cdots, x_n)^{-1} = (x_1^{-1}, x_2^{-1}, \cdots, x_n^{-1})$.

Proposition 5.10. *Let $A_1, A_2, \cdots, A_n$ be fuzzy subgroups of $G_1, G_2, \cdots, G_n$ respectively. Then the Cartesian product $\prod_{i=1}^{n} A_i = A_1 \times A_2 \times \cdots \times A_n$ is a fuzzy subgroup of $G_1 \times G_2 \times \cdots \times G_n$.*

Proof. Firstly, for any $(x_1, x_2, \cdots, x_n) \in G_1 \times G_2 \times \cdots \times G_n$,

$$(\prod_{i=1}^{n} A_i)((x_1, x_2, \cdots, x_n)^{-1}) = (\prod_{i=1}^{n} A_i)(x_1^{-1}, x_2^{-1}, \cdots, x_n^{-1})$$
$$= \bigwedge_{i=1}^{n} A_i(x_i^{-1}) = \bigwedge_{i=1}^{n} A_i(x_i)$$
$$= (\prod_{i=1}^{n} A_i)(x_1, x_2, \cdots, x_n)$$

since A_i is a fuzzy subgroup of G_i for each i ($1 \leq i \leq n$).

Next, for any $(x_1, x_2, \cdots, x_n), (y_1, y_2, \cdots, y_n) \in G_1 \times G_2 \times \cdots \times G_n$,

$$(\prod_{i=1}^{n} A_i)((x_1, x_2, \cdots, x_n)(y_1, y_2, \cdots, y_n))$$
$$= (\prod_{i=1}^{n} A_i)(x_1 y_1, x_2 y_2, \cdots, x_n y_n)$$
$$= \bigwedge_{i=1}^{n} A_i(x_i y_i) \geq \bigwedge_{i=1}^{n} (A_i(x_i) \wedge A_i(y_i))$$

$$= (\bigwedge_{i=1}^{n} A_i(x_i)) \wedge (\bigwedge_{i=1}^{n} A_i(y_i))$$

$$= (\prod_{i=1}^{n} A_i)(x_1, x_2, \cdots, x_n) \wedge (\prod_{i=1}^{n} A_i)(y_1, y_2, \cdots, y_n).$$

By definition, $\prod_{i=1}^{n} A_i$ is a fuzzy subgroup of $G_1 \times G_2 \times \cdots \times G_n$. □

5.2.2 Normal Fuzzy Subgroups

The notion of a normal fuzzy subgroup was firstly introduced by Wu [159] in 1981, which is defined as follows.

Definition 5.6. *A fuzzy subgroup A of G is called normal if $A(xy) = A(yx)$ holds for any $x, y \in G$.*

Clearly, if G is commutative, then every fuzzy group A of G is normal.

Proposition 5.11. *A fuzzy subgroup A of G is normal iff $A(xyx^{-1}) = A(y)$ holds for any $x, y \in G$.*

Proof. Suppose A is normal. By definition, for any $x, y \in G$, $A(xyx^{-1}) = A(yx^{-1}x) = A(y)$.

Conversely, suppose $A(xyx^{-1}) = A(y)$ holds for any $x, y \in G$. Then $A(xy) = A(xyxx^{-1}) = A(yx)$, i.e. A is normal. □

It is easily verified that the equality $A(xyx^{-1}) = A(y)$ in Proposition 5.11 can be replaced by $A(xyx^{-1}) \geq A(y)$ or $A(xyx^{-1}) \leq A(y)$. Indeed, if $A(xyx^{-1}) \geq A(y)$ holds for any $x, y \in G$, then

$$A(xyx^{-1}) \leq A(x^{-1}xyx^{-1}(x^{-1})^{-1}) = A(y).$$

Hence $A(xyx^{-1}) = A(y)$. Similarly, $A(xyx^{-1}) \leq A(y)$ implies $A(xyx^{-1}) = A(y)$ as well.

Proposition 5.12. *$A \in F(G)$ is a normal fuzzy subgroup of G iff $A \circ A^{-1} = A$ and $A \circ B = B \circ A$ holds for all $B \in F(G)$.*

Proof. If A is a normal fuzzy subgroup of G, then $A \circ A^{-1} = A$ by Proposition 5.7. In addition, $\forall z \in G, \forall B \in F(G)$,

$$(A \circ B)(z) = \bigvee_{xy=z} (A(x) \wedge B(y))$$

$$= \bigvee_{y \in G} (A(zy^{-1}) \wedge B(y))$$

$$= \bigvee_{y \in G} (A(y^{-1}z) \wedge B(y)) \quad (A \text{ is normal})$$

$$= \bigvee_{yx=z} (B(y) \wedge A(x))$$

$$= (B \circ A)(z).$$

Hence $A \circ B = B \circ A$.

Conversely, suppose $A \circ A^{-1} = A$ and $A \circ B = B \circ A$ holds for all $B \in F(G)$. Then A is a fuzzy subgroup of G by Proposition 5.7. For any $x \in G$, $A \circ \{x^{-1}\} = \{x^{-1}\} \circ A$. Hence, $\forall y \in G$, $(A \circ \{x^{-1}\})(y) = (\{x^{-1}\} \circ A)(y)$. Since $(A \circ \{x^{-1}\})(y) = \bigvee_{st=y} (A(s) \wedge \{x^{-1}\}(t)) = A(yx)$ and $(\{x^{-1}\} \circ A)(y) = A(xy)$, we have $A(xy) = A(yx)$, and thus A is normal. □

Proposition 5.13. $A \in F(G)$ *is a normal fuzzy subgroup of* G *iff* A_α *is a normal subgroup of* G *for any* $\alpha \in rng(A)$.

Proof. Suppose A is a normal fuzzy subgroup of G and $\alpha \in rng(A)$. Then A_α is a subgroup of G by Proposition 5.6. Let $x \in G$ and $y \in A_\alpha$. It follows from Proposition 5.11 that $A(xyx^{-1}) = A(y) \geq \alpha$. Hence $xyx^{-1} \in A_\alpha$, and thus A_α is normal.

Conversely, suppose A_α is a normal subgroup of G for every $\alpha \in rng(A)$. It follows that A is a fuzzy group of G by Proposition 5.6. Let $x, y \in G$ and $\alpha = A(y)$. Then $\alpha \in \{A(x)|x \in G\}$ and $y \in A_\alpha$. Hence $xyx^{-1} \in A_\alpha$. Consequently, $A(xyx^{-1}) \geq \alpha = A(y)$. As a result, A is a normal fuzzy subgroup of G. □

Particularly, A^* is a normal subgroup of G if A is a normal fuzzy subgroup of G.

Definition 5.7. *Let* A *be a fuzzy subgroup of* G. *For every* $x \in G$, *define* $xA, Ax \in F(G)$ *by:*

$$\forall y \in G, (xA)(y) = A(x^{-1}y) \text{ and } (Ax)(y) = A(yx^{-1}).$$

Then xA *and* Ax *are called the left coset and right coset of* A *w.r.t.* x *respectively.*

Clearly, $xA = Ax$ holds for any $x \in G$ if A is a normal fuzzy subgroup of G. In this case, we simply call $xA(= Ax)$ a coset. Write $G/A = \{xA|x \in G\}$.

Lemma 5.1. *Let* A *be two normal fuzzy subgroups of* G. *Then* $xA \circ yA = (xy)A$ *holds for any two cosets* $xA, yA \in G/A$.

Proof. On the one hand, for any $z \in G$,

$$(xA \circ yA)(z) = \bigvee_{z_1 z_2 = z} ((xA)(z_1) \wedge (yA)(z_2))$$

$$\geq (xA)(x) \wedge (yA)(x^{-1}z)$$

$$= A(x^{-1}x) \wedge A(y^{-1}x^{-1}z)$$

$$= A(e) \wedge A(y^{-1}x^{-1}z)$$
$$= A((xy)^{-1}z) = ((xy)A)(z).$$

On the other hand, considering that A is normal,

$$\begin{aligned}
(xA \circ yA)(z) &= \bigvee_{z_1 z_2 = z} ((xA)(z_1) \wedge (yA)(z_2)) \\
&= \bigvee_{z_1 z_2 = z} (A(x^{-1}z_1) \wedge A(y^{-1}z_2)) \\
&= \bigvee_{z_1 z_2 = z} (A(x^{-1}z_1) \wedge A(z_2 y^{-1})) \\
&\leq \bigvee_{z_1 z_2 = z} A(x^{-1}z_1 z_2 y^{-1}) \\
&= A(x^{-1}zy^{-1}) = A(y^{-1}x^{-1}z) \\
&= A((xy)^{-1}z) = ((xy)A)(z).
\end{aligned}$$

Hence $xA \circ yA = (xy)A$. □

We have the following result concerning $(G/A, \circ)$.

Proposition 5.14. *Let A be a normal fuzzy subgroup of G. Then*

(1) $(G/A, \circ)$ is a group and
(2) G/A is isomorphic to G/A^*.

Proof. (1) Clearly, the operation $\circ$ is associative, A is the identity of G/A and the inverse of xA is $x^{-1}A$. Hence $(G/A, \circ)$ is a group.

(2) For any $x \in G$, let $f : xA \to xA^*$. Then, for any $x, y \in G$,

$$f(xA \circ yA) = f(xyA) = xyA^* = xA^*yA^* = f(xA)f(yA).$$

Hence f is a homomorphism. In order to prove that f is injective, suppose that $xA = yA$. Then $A(x^{-1}z) = A(y^{-1}z)$ for all $z \in G$. Particularly, $A(x^{-1}y) = A(e)$ when $z = y$. Thus $x^{-1}y \in A^*$. As a result, $xA^* = yA^*$. Hence f is injective. It is clear that f is surjective. In summary, f is an isomorphism between G/A and G/A^*. □

G/A will be called the quotient group of G by a normal fuzzy subgroup A of G.

Proposition 5.15. *Let A be a normal fuzzy subgroup of G. Define $\bar{A} : G/A \to [0,1]$ by:*
$$\forall xA \in G/A, \bar{A}(xA) = A(x).$$
Then $\bar{A}$ is a normal fuzzy subgroup of G/A.

Proof. Firstly, for any $xA \in G/A$,

$$\bar{A}((xA)^{-1}) = \bar{A}(x^{-1}A) = A(x^{-1}) = A(x) = \bar{A}(xA)$$

and for any $xA, yA \in G/A$,

$$\bar{A}(xA \circ yA) = \bar{A}(xyA) = A(xy)$$
$$\geq A(x) \wedge A(y) = \bar{A}(xA) \wedge \bar{A}(yA).$$

Hence $\bar{A}$ is a fuzzy subgroup of G/A. Next, for any $xA, yA \in G/A$,

$$\bar{A}(xA \circ yA) = \bar{A}(xyA) = A(xy) = A(yx)$$
$$= \bar{A}(yxA) = \bar{A}(yA \circ xA).$$

Hence $\bar{A}$ is a normal fuzzy subgroup of G/A. $\square$

Proposition 5.16. *Let A be a normal fuzzy subgroup of G and let f be an epimorphism of G onto a group G'. Then $f(A)$ is a normal fuzzy subgroup of G'.*

Proof. By Proposition 5.8, $f(A)$ is a fuzzy subgroup of G'. Let $u, v \in G'$. Then there exists $x \in G$ such that $f(x) = u$ since f is surjective. Hence, we obtain successively

$$f(A)(uvu^{-1}) = \bigvee_{f(z)=uvu^{-1}} A(z) = \bigvee_{f(z)=f(x)v(f(x))^{-1}} A(z)$$
$$= \bigvee_{f(x^{-1}zx)=v} A(z) \quad (f \text{ is a homomorphism})$$
$$= \bigvee_{f(y)=v} A(xyx^{-1}) = \bigvee_{f(y)=v} A(y) \quad (A \text{ is normal})$$
$$= f(A)(v).$$

Hence $f(A)$ is a normal fuzzy subgroup of G'. $\square$

Proposition 5.17. *Let f be a homomorphism from G to a group G' and let B be a normal fuzzy subgroup of G'. Then $f^{-1}(B)$ is a normal fuzzy subgroup of G.*

Proof. By Proposition 5.9, $f^{-1}(B)$ is a fuzzy subgroup of G. Now, let $x, y \in G$. Then

$$f^{-1}(B)(xy) = B(f(xy)) = B(f(x)f(y))$$
$$= B(f(y)f(x)) = B(f(yx))$$
$$= f^{-1}(B)(yx).$$

Hence $f^{-1}(B)$ is a normal fuzzy subgroup of G. $\square$

Proposition 5.18. Let $A_1, A_2, \cdots, A_n$ be normal fuzzy subgroups of G_1, $G_2, \cdots, G_n$ respectively. Then the Cartesian product $\prod_{i=1}^{n} A_i$ is a normal fuzzy subgroup of $G_1 \times G_2 \times \cdots \times G_n$.

Proof. By Proposition 5.10, $\prod_{i=1}^{n} A_i$ is a fuzzy subgroup of $G_1 \times G_2 \times \cdots \times G_n$. Furthermore, $\forall (x_1, x_2, \cdots, x_n), (y_1, y_2, \cdots, y_n) \in G_1 \times G_2 \times \cdots \times G_n$,

$$(\prod_{i=1}^{n} A_i)((x_1, x_2, \cdots, x_n)(y_1, y_2, \cdots, y_n))$$
$$= (\prod_{i=1}^{n} A_i)(x_1 y_1, x_2 y_2, \cdots, x_n y_n)$$
$$= \bigwedge_{i=1}^{n} A_i(x_i y_i) = \bigwedge_{i=1}^{n} A_i(y_i x_i)$$
$$= (\prod_{i=1}^{n} A_i)((y_1, y_2, \cdots, y_n)(x_1, x_2, \cdots, x_n)).$$

Hence, $\prod_{i=1}^{n} A_i$ is a normal fuzzy subgroup of $G_1 \times G_2 \times \cdots \times G_n$. □

5.2.3 Fuzzy Subrings

In this and next subsection, we assume $(R, +, \circ)$ is a ring. For convenience, we write xy instead of $x \circ y$ for $x, y \in R$.

Definition 5.8. $A \in F(R)$ is called a *fuzzy subring* of R if A satisfies that (1) $\forall x, y \in R$, $A(x - y) \geq A(x) \wedge A(y)$ and (2) $\forall x, y \in R$, $A(xy) \geq A(x) \wedge A(y)$.

From the definition, it follows that A is a fuzzy subgroup of R under the addition $+$ if A is a fuzzy subring of R. Furthermore, this fuzzy subgroup is normal since the addition is commutative. As a result, $\forall x \in R$, $A(x) \leq A(0)$ for every fuzzy subring A, where 0 denotes the zero element of R.

Proposition 5.19. $A \in F(R)$ is a fuzzy subring of R iff A_α is a subring of R for every $\alpha \in rng(A)$.

The proof is similar to that of Proposition 5.6.

By Proposition 5.19, $A^* = \{x | A(x) = A(0)\}$ is a subring of R.

The operations on R can be extended to $F(R)$ as follows: $\forall A, B \in F(R)$, $\forall z \in R$,
$$(A + B)(z) = \bigvee_{x+y=z} (A(x) \wedge B(y));$$

$$(A - B)(z) = \bigvee_{x-y=z} (A(x) \wedge B(y));$$

$$(A \circ B)(z) = \bigvee_{xy=z} (A(x) \wedge B(y)).$$

Proposition 5.20. $A \in F(R)$ *is a fuzzy subring of R iff $A - A \subseteq A$ and $A \circ A \subseteq A$.*

Proof. Let A be a fuzzy subring of R. Since A is a fuzzy group under addition, $A - A \subseteq A$ by Proposition 5.7. Moreover, $\forall z \in R$,

$$(A \circ A)(z) = \bigvee_{z=xy} (A(x) \wedge A(y)) \leq \bigvee_{z=xy} A(xy) = A(z),$$

i.e. $A \circ A \subseteq A$.

Conversely, suppose that $A - A \subseteq A$ and $A \circ A \subseteq A$. Then, $\forall x, y \in R$,

$$A(x - y) \geq (A - A)(x - y) = \bigvee_{s-t=x-y} (A(s) \wedge A(t)) \geq A(x) \wedge A(y).$$

Similarly,

$$A(xy) \geq (A \circ A)(xy) = \bigvee_{st=xy} (A(s) \wedge A(t)) \geq A(x) \wedge A(y).$$

Consequently, A is a fuzzy subring of R. □

Proposition 5.21. *Let A be a fuzzy subring of R and let f be an epimorphism of R onto a ring R'. Then $f(A)$ is a fuzzy subring of R'.*

Proof. Let $u, v \in R'$. Then there exist $x, y \in R$ such that $f(x) = u$ and $f(y) = v$ since f is surjective. Hence, we obtain successively

$$f(A)(u) \wedge f(A)(v) = \left(\bigvee_{f(x)=u} A(x) \right) \wedge \left(\bigvee_{f(y)=v} A(y) \right)$$

$$= \bigvee_{f(x)=u, f(y)=v} (A(x) \wedge A(y))$$

$$\leq \bigvee_{f(x)=u, f(y)=v} A(x - y) \quad (A \text{ is a fuzzy subring of } R)$$

$$\leq \bigvee_{f(x)-f(y)=u-v} A(x - y)$$

$$= \bigvee_{f(x-y)=u-v} A(x - y) \quad (f \text{ is a homomorphism})$$

5.2 Fuzzy Algebra

$$= \bigvee_{f(z)=u-v} A(z)$$
$$= f(A)(u-v).$$

Similarly,
$$f(A)(uv) \geq f(A)(u) \wedge f(A)(v).$$

Hence, $f(A)$ is a fuzzy subring of R'. □

Proposition 5.22. *Let f be a homomorphism from R to a ring R' and let B be a fuzzy subring of R'. Then $f^{-1}(B)$ is a fuzzy subring of R.*

Proof. For any $x, y \in R$, $f^{-1}(B)(xy) = B(f(xy)) = B(f(x)f(y)) \geq B(f(x)) \wedge B(f(y)) = f^{-1}(B)(x) \wedge f^{-1}(B)(y)$. Similarly, $f^{-1}(B)(x-y) \geq f^{-1}(B)(x) \wedge f^{-1}(B)(y)$. Thus $f^{-1}(B)$ is a fuzzy subring of R. □

5.2.4 Fuzzy Ideals

Definition 5.9. *A fuzzy subring A of R is called a fuzzy ideal of R if it satisfies that, for any $x, y \in R$,*
$$A(xy) \geq A(x) \vee A(y).$$

Clearly, $A \in F(R)$ is a fuzzy ideal of R iff A satisfies that, $\forall x, y \in R$, $A(x-y) \geq A(x) \wedge A(y)$ and $A(xy) \geq A(x) \vee A(y)$. If R is commutative, then a fuzzy subring A of R is a fuzzy ideal iff R satisfies that, for any $x, y \in R$, $A(xy) \geq A(x)$.

Proposition 5.23. *Let $A \in F(R)$. Then A is a fuzzy ideal of R iff A_α is an ideal of R for every $\alpha \in rng(A)$.*

Proof. Firstly, suppose that A is a fuzzy ideal of R. By Proposition 5.19, A_α ($\alpha \in rng(A)$) is a subring of R. Let $x, y \in A_\alpha$ and $z \in R$. Then $A(x-y) \geq A(x) \wedge A(y) \geq \alpha$ and $A(zx) \geq A(z) \vee A(x) \geq A(x) \geq \alpha$. Hence, $x - y \in A_\alpha$ and $zx \in A_\alpha$. Thus A_α is an ideal of R.

Conversely, suppose that A_α is an ideal of R for every $\alpha \in rng(A)$. By Proposition 5.19, A is a fuzzy subring of R. Let $x, y \in R$ and $\alpha = A(x)$. Then $\alpha \in rng(A)$ and $x \in A_\alpha$. Since A_α is an ideal, $xy \in A_\alpha$. Hence $A(xy) \geq \alpha = A(x)$. Similarly, $A(xy) \geq A(y)$. Therefore, $A(xy) \geq A(x) \vee A(y)$. Thus A is a fuzzy ideal of R. □

Particularly, $A^* = \{x | A(x) = A(0)\}$ is an ideal of R if A is a fuzzy ideal of R.

Proposition 5.24. *Let A be a fuzzy ideal of R and let f be an epimorphism of R onto a ring R'. Then $f(A)$ is a fuzzy ideal of R'.*

Proof. $f(A)$ is a fuzzy ideal of R' by Proposition 5.21. Let $u, v \in R'$. Then there exist some $x, y \in R$ such that $f(x) = u$ and $f(y) = v$ since f is surjective. Hence, we obtain

$$f(A)(uv) = \bigvee_{f(z)=uv} A(z) = \bigvee_{f(z)=f(x)f(y)} A(z)$$
$$= \bigvee_{f(z)=f(xy)} A(z) \geq A(xy)$$
$$\geq A(x) \vee A(y).$$

Therefore,

$$f(A)(uv) \geq \bigvee_{f(x)=u, f(y)=v} (A(x) \vee A(y))$$
$$= \left(\bigvee_{f(x)=u} A(x)\right) \vee \left(\bigvee_{f(y)=v} A(y)\right)$$
$$= f(A)(u) \vee f(A)(v).$$

Thus $f(A)$ is a fuzzy ideal of R'. □

Proposition 5.25. *Let f be a homomorphism from R to a ring R' and let B be a fuzzy subring of R'. Then $f^{-1}(B)$ is a fuzzy subring of R.*

Proof. $f^{-1}(B)$ is a fuzzy subring of R by Proposition 5.22. For any $x, y \in R$, $f^{-1}(B)(xy) = B(f(xy)) = B(f(x)f(y)) \geq B(f(x)) \vee B(f(y)) = f^{-1}(B)(x) \vee f^{-1}(B)(y)$. Thus $f^{-1}(B)$ is a fuzzy ideal of R. □

Let R be a fuzzy ideal and R/A denote the quotient group of R by A under the addition of R, i.e. $R/A = \{x + A | x \in R\}$. By Lemma 5.1, $(x + A) + (y + A) = (x + y) + A$ for any $x + A, y + A \in R/A$. In addition, define $(x + A) \cdot (y + A) = xy + A$.

Proposition 5.26. *Let A be a fuzzy ideal of R. The above operation $\cdot$ on R/A is well-defined.*

Proof. Let $x_1 + A = x_2 + A$ and $y_1 + A = y_2 + A$. Then $\forall z \in R$, $(x_1 + A)(z) = (x_2 + A)(z)$, i.e. $A(z - x_1) = A(z - x_2)$. Particularly, when $z = x_1$, $A(x_1 - x_2) = A(0)$. Similarly, $A(y_1 - y_2) = A(0)$. Hence,

$$A(x_1 y_1 - x_2 y_2) = A(x_1 y_1 - x_1 y_2 + x_1 y_2 - x_2 y_2)$$
$$= A(x_1(y_1 - y_2) + (x_1 - x_2)y_2)$$
$$\geq A(x_1(y_1 - y_2)) \wedge A((x_1 - x_2)y_2)$$
$$\geq (A(x_1) \vee A(y_1 - y_2)) \wedge (A(x_1 - x_2) \vee A(y_2))$$
$$= (A(x_1) \vee A(0)) \wedge (A(0) \vee A(y_2))$$
$$= A(0) \wedge A(0) = A(0),$$

i.e. $A(x_1y_1 - x_2y_2) = A(0)$. Thus

$$\begin{aligned}(x_2y_2 + A)(z) &= A(z - x_2y_2) \\ &= A((z - x_1y_1) + (x_1y_1 - x_2y_2)) \\ &\geq A(z - x_1y_1) \wedge A(x_1y_1 - x_2y_2) \\ &\geq A(z - x_1y_1) \wedge A(0) \\ &= A(z - x_1y_1) = (x_1y_1 + A)(z),\end{aligned}$$

whence $x_2y_2 + A \supseteq x_1y_1 + A$. Similarly, $x_1y_1 + A \supseteq x_2y_2 + A$. Hence, $x_1y_1 + A = x_2y_2 + A$. □

Proposition 5.27. *Let A be a fuzzy ideal of R. Then $(R/A, +, \cdot)$ is a ring.*
The proof is straightforward which is left to the reader as an exercise.

$(R/A, +, \cdot)$ is called the quotient ring of R by the ideal A.

Remark 5.2. *The definition of fuzzy subgroup was introduced by Rosenfeld [104] and was extended to T-fuzzy subgroup by Anthony and Sherwood [5] and to L-fuzzy subgroup by Negoita and Ralescu [108]. Normal fuzzy subgroups were investigated by Wu [159], Mukherjee and Bhattacharya [107], Liu [91] and others. More general definition and properties of the product of fuzzy subgroups can be found in [132]. Various isomorphism theorems can be found in [105, 169]. Liu introduced the concepts of fuzzy subring and fuzzy ideal in [91] and we recommend the references [15, 38, 43, 86, 100] for further reading on these topics. For a systematic and exhaustive summation of research results related to fuzzy algebra, see [105, 132].*

5.3 Fuzzy Topology

Undoubtedly general topology has been the first branch of mathematics that has been fuzzified or colored. Already in 1968, Chang published his seminal paper entitled "Fuzzy topological spaces" in Journal of Mathematical Analysis and its Applications [25]. This fuzzification has been the starting point for many so-called direct fuzzifications of mathematical concepts and structures that are based on set-theoretic operations such as union and intersection. Since its introduction in 1968, fuzzy topology has developed to a flourishing branch of nowadays mathematics, especially in China, Egypt and India. Many research teams are active on fuzzy topological structures. The purpose of this section consists in providing a flavor of the basic concepts and results in fuzzy topology. For more details, we refer to the literature and to the existing specialized monographs (e.g. Liu and Luo [92] and Lowen [95]).

5.3.1 Definitions

Let X be a non-empty set and $F(X)$ the class of all fuzzy sets on X. A subclass $\mathcal{T}$ of $F(X)$ is called a Chang fuzzy topology (or simply a fuzzy topology) if the following three conditions are fulfilled:

(O.1) The constant $X \to \{0\}$ and $X \to \{1\}$ mappings belong to $\mathcal{T}$;
(O.2) If $O_1 \in \mathcal{T}$ and $O_2 \in \mathcal{T}$, then $O_1 \cap O_2 \in \mathcal{T}$;
(O.3) If $(\forall i \in I)$, $O_i \in \mathcal{T}$, then $\bigcup_{i \in I} O_i \in \mathcal{T}$, where I denotes an arbitrary index set.

So a fuzzy topology on X is simply a class of fuzzy sets on X that is closed under finite (Zadeh) intersections and arbitrary (Zadeh) unions. The elements of a fuzzy topology are called open (fuzzy) sets. The couple $(X, \mathcal{T})$, where $\mathcal{T}$ is a fuzzy topology on X is called a (Chang) fuzzy topological space. It is easy to see that a fuzzy topology generalizes the concept of a classical topology since Zadeh's operations modeled by minimum and supremum coincide with classical intersections and arbitrary unions in case of a two-valued membership set. Some additional concepts for $A \in F(X)$ and $(X, \mathcal{T})$ a fuzzy topological space are:

- A is called a closed (fuzzy) set iff $A^c \in \mathcal{T}$. The class of all closed fuzzy sets is denoted by $\mathcal{T}'$, i.e. $\mathcal{T}' = \{F | F^c \in \mathcal{T}\}$.
- The interior of A, denoted $\text{int}(A)$, is defined as the largest open fuzzy set contained in A, i.e. $\text{int}(A) = \bigcup\{O | O \in \mathcal{T} \text{ and } O \subseteq A\}$.
- The closure of A, denoted $\text{cl}(A)$, is defined as the smallest closed fuzzy set that contains A, i.e. $\text{cl}(A) = \bigcap\{F | F \in \mathcal{T}' \text{ and } A \subseteq F\}$.
- A fuzzy set in X is called a fuzzy singleton iff its support is a crisp singleton. A fuzzy singleton with support $\{x\}$ and value ϵ in x will be denoted by x_ϵ. The class of fuzzy singletons in X is denoted by $\mathcal{S}(X)$.
- A fuzzy set on X is called a fuzzy point iff it is a fuzzy singleton and not a crisp singleton, i.e. $p = x_\epsilon$ with $x \in X$ and $\epsilon \in]0, 1[$. The class of fuzzy points on X is denoted by $\mathcal{P}(X)$.
- A is called a Ludescher neighborhood of $x \in X$ iff $(\exists O \in \mathcal{T})(x \in \text{supp}(O)$ and $O \subseteq A)$. The class of all Ludescher neighborhoods of x is denoted by $\mathcal{L}_x(\mathcal{T})$[97].
- A is called a Kerre neighborhood of $s \in \mathcal{S}(X)$ iff $(\exists O \in \mathcal{T})(s \subseteq O \subseteq A)$. The class of all Kerre neighborhoods of s is denoted by $\mathcal{K}_s(\mathcal{T})$[79].
- A is called a Warren neighborhood of $x \in X$ iff $(\exists O \in \mathcal{T})(x \in \text{supp}(O)$ and $O \subseteq A$ and $O(x) = A(x))$. The class of all Warren neighborhoods of x is denoted by $\mathcal{W}_x(\mathcal{T})$[157].
- A is called a Pu neighborhood of $s \in \mathcal{S}(X)$ iff $(\exists O \in \mathcal{T})(sqO$ and $O \subseteq A)$, where the quasi-coincidence relation q is defined by:

$$sqO \iff \neg(s \subseteq O^c),$$

or equivalently, if $s = x_\epsilon$, $sqO \iff O(x) + \epsilon > 1$. The class of all Pu neighborhoods of s is denoted by $\mathcal{P}_s(\mathcal{T})$[117].

- A is called a Mashhour neighborhood of $p \in \mathcal{P}(X)$ iff $(\exists O \in \mathcal{T})$ $(p \lessgtr O$ and $O \subseteq A)$, where $p \lessgtr O$ is defined as $\epsilon < O(x)$ for $p = x_\epsilon$. The class of all Mashhour neighborhoods of p is denoted by $\mathcal{M}_p(\mathcal{T})$[61].

- A subset $\mathcal{B}$ of a fuzzy topology $\mathcal{T}$ is called a base for $\mathcal{T}$ iff each member of $\mathcal{T}$ can be written as a (Zadeh) union of elements of $\mathcal{B}$.
- A subset S of a fuzzy topology $\mathcal{T}$ is called a subbase for $\mathcal{T}$ iff finite (Zadeh) intersections of elements of S constitute a base for $\mathcal{T}$.
- A Chang fuzzy topological space $(X, \mathcal{T})$ is called a Lowen or stratified fuzzy topological space [94] iff the following supplementary condition is satisfied:

$$(\forall k \in [0,1])(\underline{k} \in \mathcal{T}),$$

where $\underline{k}$ denotes the constant $X - \{k\}$ mapping, or equivalently,

$$(\forall k \in [0,1])(\exists O \in \mathcal{T})(\forall x \in X)(O(x) = k).$$

- A Chang fuzzy topological space $(X, \mathcal{T})$ is called a surjective fuzzy topological space [77] iff

$$(\forall k \in [0,1])(\forall x \in X)(\exists O \in \mathcal{T})(O(x) = k).$$

Note that a Lowen fuzzy topological space may be considered as a kind of uniform surjective fuzzy topological space.

- A Chang fuzzy topological space $(X, \mathcal{T})$ is called α-generated ($\alpha \in [0, 1[$) [78] iff ($\exists T, T$ is a crisp topology on X) ($\mathcal{T} = C(X, [0,1]_\alpha)$), i.e. $\mathcal{T}$ consists of all continuous mappings between the topological spaces (X, T) and $([0,1], \{\emptyset, [0,1],]\alpha, 1]\})$.
- A Chang fuzzy topological space $(X, \mathcal{T})$ is called topologically generated iff

($\exists T, T$ is a crisp topology on X) ($\mathcal{T} = C(X, [0,1])$),

i.e. $\mathcal{T}$ consists of all continuous mappings between the topological spaces (X, T) and $([0,1], \{\emptyset, [0,1], \{]\alpha, 1]|\alpha \in [0, 1[\}\})$.
- A Chang fuzzy topological space $(X, \mathcal{T})$ is called a Morsi fuzzy topological space iff

$$(\forall \alpha \in [0,1[)(\forall O \in \mathcal{T})(O^\alpha \in \mathcal{T}),$$

where O^α is defined as the fuzzy set on X:

$$O^\alpha(x) = \begin{cases} O(x) & \text{if } O(x) > \alpha \\ 0 & \text{elsewhere} \end{cases}.$$

5.3.2 Characterization of a Fuzzy Topology in Terms of Preassigned Operations

First of all, we shall establish the interrelations between fuzzy points and fuzzy singletons on the one hand and the inclusion, equality and elementary set-theoretic operations on the other hand in terms of the three possible "membership relations" $\subseteq$, $\leqslant$ and q.

Proposition 5.28. *Let s be a fuzzy singleton in X, A and B fuzzy sets on X and $\{A_j\}(j \in J)$ a family of fuzzy sets on X. Then the following properties hold:*

$A \subseteq B \iff (\forall s \in \mathcal{S}(X))(s \subseteq A \Rightarrow s \subseteq B)$
$A = B \iff (\forall s \in \mathcal{S}(X))(s \subseteq A \Leftrightarrow s \subseteq B)$
$s \subseteq \bigcap_{j \in J} A_j \iff (\forall j \in J)(s \subseteq A_j)$
$s \subseteq \bigcup_{j \in J} A_j \Leftarrow (\exists j \in J)(s \subseteq A_j)$
$s \subseteq \bigcup_{j=1}^{n} A_j \iff (\exists j \in \{1, 2, \cdots, n\})(s \subseteq A_j)$
$A = \bigcup_{s \subseteq A} s.$

Proposition 5.29. *Let p be a fuzzy point in X, A and B fuzzy sets on X and $\{A_j\}(j \in J)$ a family of fuzzy sets on X. Then the following properties hold:*

$A \subseteq B \iff (\forall p \in \mathcal{P}(X))(p \mathrel{\underset{\sim}{\in}} A \Rightarrow p \mathrel{\underset{\sim}{\in}} B)$
$A = B \iff (\forall p \in \mathcal{P}(X))(p \mathrel{\underset{\sim}{\in}} A \Leftrightarrow p \mathrel{\underset{\sim}{\in}} B)$
$p \mathrel{\underset{\sim}{\in}} \bigcap_{j \in J} A_j \Rightarrow (\forall j \in J)(p \mathrel{\underset{\sim}{\in}} A_j)$
$p \mathrel{\underset{\sim}{\in}} \bigcap_{j=1}^{n} A_j \iff (\forall j \in \{1, 2, \cdots, n\})(p \mathrel{\underset{\sim}{\in}} A_j)$
$p \mathrel{\underset{\sim}{\in}} \bigcup_{j \in J} A_j \iff (\exists j \in J)(p \mathrel{\underset{\sim}{\in}} A_j)$
$A = \bigcup_{p \mathrel{\underset{\sim}{\in}} A} p.$

Proposition 5.30. *Let s be a fuzzy singleton in X, A and B fuzzy sets on X and $\{A_j\}(j \in J)$ a family of fuzzy sets on X. Then the following properties hold:*

$A \subseteq B \iff (\forall s \in \mathcal{S}(X))(sqA \Rightarrow sqB)$
$A = B \iff (\forall s \in \mathcal{S}(X))(sqA \Leftrightarrow sqB)$
$sq \bigcap_{j \in J} A_j \Rightarrow (\forall j \in J)(sqA_j)$
$sq \bigcap_{j=1}^{n} A_j \iff (\forall j \in \{1, 2, \cdots, n\})(sqA_j)$
$sq \bigcup_{j \in J} A_j \iff (\exists j \in J)(sqA_j)$
$A = \bigcup_{\neg(sqA^c)} s.$

Proof. As an example, we prove the third formula. By putting $s = x_\epsilon$, we get successively:

5.3 Fuzzy Topology

$$sq \bigcup_{j \in J} A_j \Leftrightarrow \epsilon + (\bigcup_{j \in J} A_j)(x) > 1$$
$$\Longleftrightarrow \epsilon + \sup_{j \in J} A_j(x) > 1$$
$$\Longleftrightarrow \sup_{j \in J}(\epsilon + A_j(x)) > 1$$
$$\Longleftrightarrow (\exists j \in J)(\epsilon + A_j(x) > 1)$$
$$\Longleftrightarrow (\exists j \in J)(sqA_j). \qquad \square$$

Remark 5.3. *Generally, the following generalizations of classical set theory do not hold:*

$s \subseteq A^c \Longleftrightarrow \neg(s \subseteq A)$
$p \not\in A^c \Longleftrightarrow \neg(p \not\in A)$
$sqA^c \Longleftrightarrow \neg(sqA)$ *for* $A \in F(X)$, $s \in \mathcal{S}(X)$ *and* $p \in \mathcal{P}(X)$.

Indeed, putting $X = \{x_0\}$, $A(x_0) = 0.5$ and $s(x_0) = \frac{2}{3}$, we get: $s \not\subseteq A$ and $s \not\subseteq A^c$, i.e.

$$\neg(\forall X)(\forall s \in \mathcal{S}(X))(\forall A \in F(X))(s \not\subseteq A \Rightarrow s \subseteq A^c).$$

5.3.3 Characterization of a Fuzzy Topology in Terms of Closed Sets

Let $(X, \mathcal{T})$ be a fuzzy topological space and $\mathcal{T}'$ the corresponding class of closed sets. Then it is easy to show that $\mathcal{T}'$ satisfies the following properties:

(C1) $\emptyset \in \mathcal{T}'$ and $X \in \mathcal{T}'$;
(C2) $F_1 \in \mathcal{T}'$ and $F_2 \in \mathcal{T}' \Rightarrow F_1 \cup F_2 \in \mathcal{T}'$;
(C3) $(\forall j \in J)(F_j \in \mathcal{T}') \Rightarrow \bigcap_{j \in J} F_j \in \mathcal{T}'$,

i.e. the class of closed fuzzy sets is closed under finite unions and arbitrary intersections.

Conversely, a fuzzy topology is completely determined by its "closed" fuzzy sets. More precisely, let $X \neq \emptyset$ and $\mathcal{T}^* \subseteq F(X)$ satisfying:

(C1') $\emptyset \in \mathcal{T}^*$ and $X \in \mathcal{T}^*$;
(C2') $F_1 \in \mathcal{T}^*$ and $F_2 \in \mathcal{T}^* \Rightarrow F_1 \cup F_2 \in \mathcal{T}^*$;
(C3') $(\forall j \in J)(F_j \in \mathcal{T}^*) \Rightarrow \bigcap_{j \in J} F_j \in \mathcal{T}^*$.

Define the class $\mathcal{T}$ as: $\mathcal{T} = \{O | O \in F(X)$ and $O^c \in \mathcal{T}^*\}$. Then $\mathcal{T}$ is a fuzzy topology on X such that the corresponding class $\mathcal{T}'$ of closed fuzzy sets equals $\mathcal{T}^*$. Indeed,

(O.1) From $\emptyset^c = X \in \mathcal{T}^*$, we obtain $\emptyset \in \mathcal{T}$. From $X^c = \emptyset \in \mathcal{T}^*$, we obtain $X \in \mathcal{T}$.

(O.2) Let $O_1 \in \mathcal{T}$ and $O_2 \in \mathcal{T}$. Hence, by definition of $\mathcal{T}$, we get $O_1^c \in \mathcal{T}^*$ and $O_2^c \in \mathcal{T}^*$. Applying (C2'), we get $O_1^c \cup O_2^c \in \mathcal{T}^*$, and hence $(O_1 \cap O_2)^c \in \mathcal{T}^*$, i.e. $O_1 \cap O_2 \in \mathcal{T}$.

(O.3) Suppose $(\forall j \in J)(O_j \in \mathcal{T})$, i.e. $(\forall j \in J)(O_j^c \in \mathcal{T}^*)$. Applying (C3'), we get $\bigcap_{j \in J} O_j^c \in \mathcal{T}^*$, or equivalently, $(\bigcup_{j \in J} O_j)^c \in \mathcal{T}^*$, i.e. $\bigcup_{j \in J} O_j \in \mathcal{T}$.

Finally, we get successively:

$$F \in \mathcal{T}' \iff F^c \in \mathcal{T} \iff (F^c)^c \in \mathcal{T}^* \iff F \in \mathcal{T}^*.$$

Hence $\mathcal{T}' = \mathcal{T}^*$.

5.3.4 Characterization of a Fuzzy Topology Using the Interior Operator

First, we need the following lemma.

Lemma 5.2. *Let $(X, \mathcal{T})$ be a fuzzy topological space. Then the following properties hold:*

(1) $(\forall A \in F(X))(int(A) \in \mathcal{T})$;
(2) $O \in \mathcal{T} \iff int(O) \in \mathcal{T}$;
(3) $(\forall A \in F(X))(cl(A) \in \mathcal{T}')$;
(4) $F \in \mathcal{T}' \iff cl(F) = F$;
(5) $(\forall A \in F(X))(int(A) \subseteq A \subseteq cl(A))$;
(6) $A \subseteq B \Rightarrow int(A) \subseteq int(B)$;
(7) $A \subseteq B \Rightarrow cl(A) \subseteq cl(B)$.

Proof. As an example, we prove (2). From $int(O) = O$ and $O \in \mathcal{T}$ we get $int(O) \in \mathcal{T}$.

Conversely, suppose $O \in \mathcal{T}$. From the definition of int, we get $int(O) \subseteq O$. From $O \in \mathcal{T}$, we get $O \in \mathcal{T}$ and $O \subseteq O$, and hence $O \in \{A | A \in \mathcal{T} \text{ and } A \subseteq O\}$, i.e. $O \subseteq int(O)$. □

Proposition 5.31. *The interior operator int in a fuzzy topological space satisfies the following properties:*

(I1) $int(X) = X$;
(I2) $(\forall A \in F(X))(int(A) \subseteq A)$;
(I3) $(\forall A \in F(X))(int(int(A)) = A)$;
(I4) $(\forall A, B \in F(X))(int(A \cap B) = int(A) \cap int(B))$.

Proof. Property (I1) follows from $X \in \mathcal{T}$ and Property (2) of Lemma 5.2. Property (I2) follows from Property (5) of Lemma 5.2. Property (I3) follows from Property (1) and (2) of Lemma 5.2. Finally, from $A \cap B \subseteq A$, $A \cap B \subseteq B$ and the monotonicity of int, we obtain $int(A \cap B) \subseteq int(A) \cap int(B)$. On the other hand, from $int(A) \in \mathcal{T}$ and $int(B) \in \mathcal{T}$, it follows that $int(A) \cap int(B) \in \mathcal{T}$.

5.3 Fuzzy Topology

Hence $int(A) \cap int(B)$ is an open fuzzy set that is contained in $A \cap B$. Since $int(A \cap B)$ is the largest open fuzzy set contained in $A \cap B$, we obtain $int(A) \cap int(B) \subseteq int(A \cap B)$. □

Conversely, a fuzzy topology is completely characterized by an interior operator.

Proposition 5.32. *Let $X \neq \emptyset$ and i an $F(X) - F(X)$ mapping that satisfies the following conditions:*

(i.1) $i(X) = X$;
(i.2) $(\forall A \in F(X))$, $(i(A) \subseteq A)$;
(i.3) $(\forall A \in F(X))$, $(i(i(A)) = A)$;
(i.4) $(\forall A, B \in F(X))$, $i(A \cap B) = i(A) \cap i(B)$.

Define the class $\mathcal{T}$ as: $\mathcal{T} = \{O | O \in F(X) \text{ and } i(O) = O\}$. Then $\mathcal{T}$ is a fuzzy topology on X satisfying $(\forall A \in F(X))$ $(int(A) = i(A))$.

Proof. From (i.1) and the definition of $\mathcal{T}$, we obtain $X \in \mathcal{T}$. From (i.2), it follows that $i(\emptyset) \subseteq \emptyset$, and hence $i(\emptyset) = \emptyset$, i.e. $\emptyset \in \mathcal{T}$.

To prove (O.2), we start with $O_1 \in \mathcal{T}$ and $O_2 \in \mathcal{T}$, i.e. $i(O_1) = O_1$ and $i(O_2) = O_2$ and hence from (i.4) it follows that $i(O_1 \cap O_2) = O_1 \cap O_2$, i.e. $O_1 \cap O_2 \in \mathcal{T}$.

Finally, suppose that $\{O_k\}$ ($k \in I$) is a family in $\mathcal{T}$, i.e. for every $k \in I$, $i(O_k) = O_k$ holds. From (i.2), we immediately get $i(\bigcup_{k \in I} O_k) \subseteq \bigcup_{k \in I} O_k$. To prove the reverse inclusion, we first show that each operator i satisfying (i.1–i.4) is increasing. Indeed, for fuzzy sets A and B in X, we obtain successively using (i.4):

$$A \subseteq B \iff A \cap B = A$$
$$\Rightarrow i(A \cap B) = i(A)$$
$$\iff i(A) \cap i(B) = i(A)$$
$$\iff i(A) \subseteq i(B).$$

Applying this monotonicity, we obtain from $O_j \subseteq \bigcup_{k \in I} O_k$ that $i(O_j) \subseteq i(\bigcup_{k \in I} O_k)$, $\forall j \in I$ and hence $\bigcup_{k \in I} i(O_k) \subseteq i(\bigcup_{k \in I} O_k)$, or equivalently, since $i(O_k) = O_k$ holds, $\bigcup_{k \in I} O_j \subseteq i(\bigcup_{k \in I} O_k)$.

Now we still have to prove that $i = int$. We know that $int(A)$ has been defined as the union of all open fuzzy sets included in A, i.e. as the union of all fuzzy sets satisfying $i(O) = O$. From (i.3), it follows that $i(A) \in \mathcal{T}$ and hence $int(i(A)) = i(A)$. From (i.2) and the monotonicity of int we get $int(i(A)) \subseteq int(A)$ and hence $i(A) \subseteq int(A)$. On the other hand, from (i.2) and monotonicity of i we get $i(int(A)) \subseteq i(A)$. Because $int(A)$ is open, we have $i(int(A)) = int(A)$. So $int(A) \subseteq i(A)$ and hence $int(A) = i(A)$. □

5.3.5 Characterization of a Fuzzy Topology by Means of a Closure Operator

Proposition 5.33. *The closure operator cl in a fuzzy topological space on X satisfies the following properties:*

(K1) $cl(\emptyset) = \emptyset$;
(K2) $(\forall A \in F(X))$, $A \subseteq cl(A)$;
(K3) $(\forall A \in F(X))$, $cl((cl(A)) = A)$;
(K4) $(\forall A, B \in F(X))$, $cl(A \cup B) = cl(A) \cup cl(B)$.

Now we claim that a fuzzy topology may be uniquely determined by means of a closure operator. We leave the proof as an exercise.

Proposition 5.34. *Let $X \neq \emptyset$ and k an $F(X) - F(X)$ mapping that satisfies the following conditions:*

(k1) $k(\emptyset) = \emptyset$;
(k2) $(\forall A \in F(X))$, $A \subseteq k(A)$;
(k3) $(\forall A \in F(X))$, $k((k(A)) = A$;
(k4) $(\forall A, B \in F(X))$, $k(A \cup B) = k(A) \cup k(B)$.

Define the class $\mathcal{T}$ as: $\mathcal{T} = \{F^c | F \in F(X) \text{ and } k(F) = F\}$. Then $\mathcal{T}$ is a fuzzy topology on X satisfying $(\forall A \in F(X))$ $(K(A) = k(A))$.

5.3.6 Characterization of a Fuzzy Topology by Means of Neighborhood Systems

From the previous sections it follows that as in the classical case a fuzzy topology may be characterized by means of the closed sets, the interior operator and the closure operator. In this section, we investigate if this characterization also holds in terms of neighborhood systems. We already know that different fuzzifications exist for the concept of a neighborhood of a point in classical topology. Here we mainly concentrate on the Kerre's definition. First we prove the characterization of an open fuzzy set in terms of Kerre neighborhoods, i.e. a fuzzy set is open iff it is a Kerre neighborhood of its fuzzy singletons.

Lemma 5.3. *Let $(X, \mathcal{T})$ be a fuzzy topological space and $O \in F(X)$. Then it holds:*
$$O \in \mathcal{T} \iff (\forall s \in \mathcal{S}(X))(s \subseteq O \Rightarrow O \in \mathcal{K}_s(\mathcal{T})).$$

Proof. The "$\Rightarrow$"' part follows directly from the definition of a Kerre neighborhood. Conversely, from $O \in \mathcal{K}_s(\mathcal{T})$, it follows that for each $s \subseteq O$ there exists an $O_s \in \mathcal{T}$ such that $s \subseteq O_s \subseteq O$ and hence
$$\bigcup_{s \subseteq O} s \subseteq \bigcup_{s \subseteq O} O_s \subseteq O,$$
i.e. $O = \bigcup_{s \subseteq O} O_s$ and so O is open as union of a family of open fuzzy sets. $\square$

5.3 Fuzzy Topology

Next we prove that the system of Kerre neighborhoods satisfies the straightforward fuzzification of the characteristic properties of neighborhood systems in classical topology.

Proposition 5.35. *Let $(X, \mathcal{T})$ be a fuzzy topological space. The class $\mathcal{K}_s(\mathcal{T})$ of Kerre neighborhoods of a fuzzy singleton s in X satisfies the following properties:*

(N1) $(\forall s \in \mathcal{S}(X)) (\mathcal{K}_s(\mathcal{T}) \neq \emptyset)$ and $(\forall V \in \mathcal{K}_s(\mathcal{T}))(s \subseteq V)$, i.e. each fuzzy singleton has at least one neighborhood and each neighborhood of a fuzzy singleton contains at least that singleton.

(N2) $(\forall s \in \mathcal{S}(X)) (\forall V_1, V_2 \in \mathcal{K}_s(\mathcal{T})) (V_1 \cap V_2 \in \mathcal{K}_s(\mathcal{T}))$, i.e. the intersection of two neighborhoods of a fuzzy singleton remains a neighborhood of that fuzzy singleton.

(N3)$(\forall s \in \mathcal{S}(X)) (\forall V \in \mathcal{K}_s(\mathcal{T})) (\forall W \in F(X)) (V \subseteq W \Rightarrow W \in \mathcal{K}_s(\mathcal{T}))$, i.e. a superset of a neighborhood of a fuzzy singleton is also a neighborhood of that fuzzy singleton.

(N4)$(\forall s \in \mathcal{S}(X)) (\forall V \in \mathcal{K}_s(\mathcal{T})) (\exists U \in \mathcal{K}_s(\mathcal{T})) (U \subseteq V$ and $(\forall s_1 \in \mathcal{S}(X)) (s_1 \subseteq U \Rightarrow V \in \mathcal{K}_{s_1}(\mathcal{T})))$, and more strongly,

(N4')$(\forall s \in \mathcal{S}(X)) (\forall V \in \mathcal{K}_s(\mathcal{T})) (\exists U \in \mathcal{K}_s(\mathcal{T})) (U \subseteq V$ and $(\forall s_1 \in \mathcal{S}(X)) (s_1 \subseteq U \Rightarrow U \in \mathcal{K}_{s_1}(\mathcal{T})))$, i.e. every neighborhood of a fuzzy singleton contains an open neighborhood.

Proof. (N1) follows immediately from $X \in \mathcal{T}$.

(N2) Let $s \in \mathcal{S}(X)$, $V_1 \in \mathcal{K}_s(\mathcal{T})$ and $V_2 \in \mathcal{K}_s(\mathcal{T})$. So there exist $O_1 \in \mathcal{T}$ and $O_2 \in \mathcal{T}$ such that $s \subseteq O_1 \subseteq V_1$ and $s \subseteq O_2 \subseteq V_2$ and hence $s \subseteq O_1 \cap O_2 \subseteq V_1 \cap V_2$ from which we get $V_1 \cap V_2 \in \mathcal{K}_s(\mathcal{T})$ since $O_1 \cap O_2 \in \mathcal{T}$.

(N3) Consider $V \in \mathcal{K}_s(\mathcal{T})$, $W \in F(X)$ and $V \subseteq W$. From $V \in \mathcal{K}_s(\mathcal{T})$, we obtain the existence of $O \in \mathcal{T}$ such that $s \subseteq O \subseteq V$ and hence $s \subseteq O \subseteq W$, i.e. $W \in \mathcal{K}_s(\mathcal{T})$.

(N4') From $V \in \mathcal{K}_s(\mathcal{T})$, we get $s \subseteq O \subseteq V$ for some $O \in \mathcal{T}$. Applying Lemma 5.3 leads to:

$$(\forall s_1 \in \mathcal{S}(X))(s_1 \subseteq O \Rightarrow O \in \mathcal{K}_{s_1}(\mathcal{T})).$$

Hence $O = U$ satisfies (N4').

Remark 5.4. *(1) Contrary to the classical situation, the properties (N1-N4) are not sufficient for the characterization of a fuzzy topology. More precisely, consider a mapping K from $\mathcal{S}(X)$ to $P(F(X))$ that transforms each s in $\mathcal{S}(X)$ into a class K_s of fuzzy sets on X and that satisfies the following conditions:*

(i) $(\forall s \in \mathcal{S}(X))(K_s \neq \emptyset)$.
(ii) $(\forall s \in \mathcal{S}(X))(\forall A, B \in K_s)(A \cap B \in K_s)$.
(iii) $(\forall s \in \mathcal{S}(X))(\forall A \in K_s)(\forall B \in F(X))(A \subseteq B \Rightarrow B \in K_s)$.

Then the class $= \{O | O \in F(X)$ and $(\forall s \subseteq O)(O \in K_s)\}$ satisfies properties (O.1) and (O.2) from Definition 5.3.1 but not necessarily (O.3) and hence $\mathcal{T}$

is not necessarily a fuzzy topology on X. Note that $s \subseteq \bigcup_{k \in I} O_k \Rightarrow (\exists k \in I)(s \subseteq O_k)$ holds but not the converse that is used in the proof of the classical characterization. In [79], it was proved that the class $\mathcal{T}$ only defines a base for a fuzzy topology on X. Moreover, in [79], the authors constructed a counterexample contradicting the assertion of Pu and Liu made in [117] concerning the characterization of a fuzzy topology by means of (Kerre) neighborhoods of fuzzy singletons.

(2) The neighborhood system of Pu-Liu [117] for fuzzy singletons and the quasi-coincidence relation provides a characterization for a fuzzy topology.

(3) The neighborhood system of Mashhour [1, 61] for fuzzy points and the membership relation $\lessgtr$ also provides a characterization for a fuzzy topology.

(4) All neighborhood systems except Warren's satisfy the direct fuzzifications of classical neighborhood systems (N1-N4).

(5) A Ludescher neighborhood system defines a fuzzy topology that is however not necessarily unique and hence it does not provide a characterization of a fuzzy topology.

(6) It is possible to move from one neighborhood system to another one. For more details about the transition formulas, we refer to [77].

5.3.7 Normality in Fuzzy Topological Spaces

As said before general topology has been one of the first disciplines in pure mathematics that has been fuzzified. So all important concepts such as compactness, separation axioms, connectedness, normality have been generalized to fuzzy topology. Similarly as for the set-theoretic operations there are many possible fuzzifications mostly with different properties. As an example, we briefly introduce here the concept of normality in fuzzy topological spaces.

A classical topological space (X, T) is called normal iff

$(\forall F_1, F_2 \in T')(F_1 \cap F_2 = \emptyset \Rightarrow (\exists O_1, O_2 \in T)(O_1 \cap O_2 = \emptyset$ and $F_1 \subseteq O_1$ and $F_2 \subseteq O_2)$, where T' denotes the class of closed sets.

Urysohn has proved the following equivalence: (X, T) is normal $\iff$ $(\forall O \in T)(\forall F, F \in T'$ and $F \subseteq O)(\exists V \in P(X))(F \subseteq int(V)$ and $cl(V) \subseteq O)$.

In 1975, Hutton[69] generalized the concept of normality to a fuzzy topological space $(X, \mathcal{T})$. $(X, \mathcal{T})$ is Hutton-normal $\iff$ $(\forall O \in \mathcal{T})(\forall F, F \in \mathcal{T}'$ and $F \subseteq O)(\exists V \in F(X))(F \subseteq int(V)$ and $cl(V) \subseteq O)$, i.e. Hutton took a straightforward fuzzification of Urysohn's form for normality. A few years later Kerre[76] answered the question why the original definition of normality has not been taken for fuzzification. The main reason behind is the fact that the equivalence for crisp sets F_1 and F_2:

$$F_1 \cap F_2 = \emptyset \iff F_1 \subseteq F_2^c,$$

5.3 Fuzzy Topology

no longer holds for fuzzy sets. Indeed, for fuzzy sets only $F_1 \cap F_2 = \emptyset \Rightarrow F_1 \subseteq F_2^c$ holds. It turned out that we have to choose the weaker form of disjointness in the fuzzification of the classical definition: $(X, \mathcal{T})$ is Kerre-normal $\iff (\forall F_1, F_2 \in \mathcal{T}')(F_1 \subseteq F_2^c \Rightarrow (\exists O_1, O_2 \in \mathcal{T})(O_1 \subseteq O_2^c$ and $F_1 \subseteq O_1$ and $F_2 \subseteq O_2)$.

Some related concepts are weak normality and complete normality: $(X, \mathcal{T})$ is weakly normal $\iff (\forall F_1, F_2 \in \mathcal{T}')(F_1 \cap F_2 = \emptyset \Rightarrow (\exists O_1, O_2 \in \mathcal{T})(O_1 \subseteq O_2^c$ and $F_1 \subseteq O_1$ and $F_2 \subseteq O_2)$.

$(X, \mathcal{T})$ is completely normal $\iff (\forall A_1, A_2 \in F(X))(A_1 \subseteq (cl(A_2))^c$ and $A_2 \subseteq (cl(A_1))^c \Rightarrow (\exists O_1, O_2 \in \mathcal{T})(A_1 \subseteq O_1$ and $A_2 \subseteq O_2$ and $O_1 \subseteq O_2^c)$.

In the next proposition, we outline the links between the different concepts of normality.

Proposition 5.36. *For a fuzzy topological space $(X, \mathcal{T})$, the following properties hold:*

(1) $(X, \mathcal{T})$ is (Kerre-)normal $\Rightarrow$ $(X, \mathcal{T})$ is weakly normal;
(2) $(X, \mathcal{T})$ is (Kerre-)normal $\iff$ $(X, \mathcal{T})$ is (Hutton-)normal;
(3) $(X, \mathcal{T})$ is completely normal $\Rightarrow$ $(X, \mathcal{T})$ is (Kerre-)normal.

We leave the proof as an exercise. Note that the reverse implication in (1) is not true; indeed the so-called fuzzy Sierpinski space [76] is a suitable counterexample.

5.3.8 Some Examples of Fuzzy Topological Spaces

Let X be a non-empty set.

(1) Indiscrete fuzzy topology $\mathcal{T}_1$: $\mathcal{T}_1 = \{\emptyset, X\}$. $\mathcal{T}_1$ is the smallest fuzzy topology on X.
(2) Discrete fuzzy topology $\mathcal{T}_2$: $\mathcal{T}_2 = F(X)$. $\mathcal{T}_2$ is the largest fuzzy topology on X.
(3) Induced fuzzy topology $\mathcal{T}_3$: $\mathcal{T}_3 = \{\chi_O | O \in T\}$, where χ_O denotes the characteristic function of O and T is a crisp topology on X.
(4) Fuzzy topology $\mathcal{T}_4$ generated by a class Σ of fuzzy sets on X. Let Σ satisfy $\{\emptyset, X\} \subseteq \Sigma \subseteq F(X)$. The fuzzy topology $\mathcal{T}_4$ with Σ as a subbasis is denoted $\langle \Sigma \rangle$ and called the fuzzy topology generated by Σ.
(5) Fuzzy Sierpinski space $\mathcal{T}_5$. Let X be a set with two elements, i.e. $X = \{x_1, x_2\}$ and $p = \{(x_1, 0), (x_2, y)\}$ with $y \in]0, 1]$ a fuzzy singleton in X. Then the fuzzy Sierpinski space associated with p is defined as:

$$\mathcal{T}_5 = \{\emptyset\} \cup \{O_{a,b} | a \in [0, 1] \text{ and } b \in [y, 1]\},$$

where $O_{a,b} = \{(x_1, a)(x_2, b)\}$.
(6) Included fuzzy singleton fuzzy topology $\mathcal{T}_6$. Let s be a fuzzy singleton in X. The class $\mathcal{T}_6$ defined as:

$$\mathcal{T}_6 = \{O | O \in F(X) \text{ and } (O = \emptyset \text{ or } s \subseteq O)\}$$

is called the included fuzzy singleton fuzzy topology on X.

As an example, we will prove that T_6 constitutes a fuzzy topology on X.

(O.1) $\emptyset \in T_6$ by the definition of T_6. $X \in T_6$ since $s \subseteq X$.
(O.2) Let $O_1, O_2 \in T_6$. The law of the excluded middle leads to $(O_1 = \emptyset$ or $O_2 = \emptyset)$ or $(O_1 \neq \emptyset$ and $O_2 \neq \emptyset)$. If $O_1 = \emptyset$ or $O_2 = \emptyset$, then $O_1 \cap O_2 = \emptyset$ and hence $O_1 \cap O_2 \in T_6$. If on the contrary $O_1 \neq \emptyset$ and $O_2 \neq \emptyset$, then by the definition of T_6, $s \subseteq O_1$ and $s \subseteq O_2$ and hence applying Proposition 5.28 yields $s \subseteq O_1 \cap O_2$, i.e. $O_1 \cap O_2 \in T_6$.
(O.3) Suppose $\{O_k\}(k \in I)$ is a family in T_6. If $(\forall k \in I)(O_k = \emptyset)$, then $\bigcup_{k \in I} O_k \in T_6$. If on the contrary $(\exists k \in I)(O_k \neq \emptyset)$, then $s \subseteq O_k$ and hence, applying again Proposition 5.28, $s \subseteq \bigcup_{k \in I} O_k$, i.e. $\bigcup_{k \in I} O_k \in T_6$.

(7) Excluded fuzzy singleton fuzzy topology T_7. Let s be a fuzzy singleton in X. The class T_7 defined as:

$$T_7 = \{O | O \in F(X) \text{ and } (O = X \text{ or } s \subseteq O^c)\}$$

is called the excluded fuzzy singleton fuzzy topology on X.

(8) Included fuzzy set fuzzy topology T_8. Let $A \in F(X)$. The class T_8 defined as:

$$T_8 = \{O | O \in F(X) \text{ and } (O = \emptyset \text{ or } A \subseteq O)\}$$

is called the included fuzzy set fuzzy topology on X.

(9) Excluded fuzzy set fuzzy topology T_9. Let $A \in F(X)$. The class T_9 defined as:

$$T_9 = \{O | O \in F(X) \text{ and } (O = X \text{ or } A \subseteq O^c)\}$$

is called the excluded fuzzy set fuzzy topology on X.

(10) Co-countable fuzzy topology T_{10}. Suppose the universe X is uncountable. The class T_{10} defined as:

$$T_{10} = \{O | O \in F(X) \text{ and } (O = \emptyset \text{ or } supp(O^c) \text{ is countable})\}$$

is called the co-countable fuzzy topology on X.

(11) The Z-fuzzy topology T_{11}. The class T_{11} defined as:

$$T_{11} = \{\emptyset, [0,1], \{(x,x) | x \in [0,1]\}\}$$

is called the Z-fuzzy topology (on $[0,1]$).

5.4 Exercises

1. Compute $\int_E A dg$ for $A \in \mathcal{A}$.

5.4 Exercises

2. If $g : \mathcal{A} \to [0, 1]$ satisfies $g(A \cup B) = g(A)$ whenever $A, B \in \mathcal{A}$, $g(B) = 0$ and $A \cap B = \emptyset$, then g is called null-additive, show that the following statements are equivalent provided that $(X, \mathcal{A}, g)$ is a fuzzy measure space:

   (1) g is null-additive;
   (2) $g(A \cup B) = g(A)$ whenever $A, B \in \mathcal{A}$ and $g(B) = 0$;
   (3) $g(A \setminus B) = g(A)$ whenever $A, B \in \mathcal{A}$ and $g(B) = 0$;
   (4) $g(A \triangle B) = g(A)$ whenever $A, B \in \mathcal{A}$ and $g(B) = 0$.

3. If s_1 and s_2 are simple functions such that $s_1 \subseteq s_2$, show that $H_A(s_1) \leq H_A(s_2)$.

4. Show that $H_A(s) = \int_A s dg$.

5. If h_1 and h_2 are two measurable functions on $(X, \mathcal{A})$ satisfying $|h_1 - h_2| < c (c \in [0, 1])$, show that

$$\left| \int_A h_1 dg - \int_A h_2 dg \right| \leq c.$$

6. Show that $\int_A h dg = \alpha$ ($\alpha \in [0, 1]$) iff $g(A \cap h_\alpha) \geq \alpha \geq g(A \cap h_{\overline{\alpha}})$.

7. If g is a fuzzy measure which is sub-additive, i.e. $g(A \cup B) \leq g(A) + g(B)$ whenever $A, B \in \mathcal{A}$ and if h_n and h are measurable functions on $(X, \mathcal{A})$ satisfying $\lim_{n \to \infty} h_n(x) = h(x)$, show that

$$\lim_{n \to \infty} \int_A h_n dg = \int_A h dg.$$

8. If $h_1 = h_2$ a.e., show that $\int_A h_1 dg = \int_A h_2 dg$ iff g is null-additive.

9. Let A be a fuzzy subgroup of G. Show that, $\forall x, y \in G$, $A(xy) = A(x) \wedge A(y)$ if $A(x) \neq A(y)$.

10. If A is a fuzzy subgroup of G, show that supp(A) is a subgroup of G.

11. Let G be a finite group. Show that A is a fuzzy subgroup of G iff $A(xy) \geq A(x) \wedge A(y)$ for any $x, y \in G$.

12. Let A be a fuzzy subgroup of G and $x \in G$. Show that, $A(xy) = A(y)$ holds for any $y \in G$ iff $A(x) = A(e)$.

13. Let A be a normal fuzzy subgroup of G and $x, y \in G$. If $xA = yA$, show that $A(x) = A(y)$.

14. If A is a fuzzy subgroup of G such that, $\forall x \in G$, $xA = Ax$, show that A is a normal fuzzy subgroup of G.

15. Let A_i ($i \in I$) be a family of fuzzy subgroups of G. Show that $\bigcap_{i \in I} A_i$ is a fuzzy subgroup of G. If, furthermore, A_i ($i \in I$) is a family of normal fuzzy subgroups of G, show that $\bigcap_{i \in I} A_i$ is a normal fuzzy subgroup of G.

16. Let A be a normal fuzzy subgroup of G and B a fuzzy subgroup of G. Show that AB is a fuzzy subgroup of G.

17. Let A and B be two normal fuzzy subgroups of G. Show that AB is a normal fuzzy subgroup of G.

18. Let A be a fuzzy subgroup of G. Set $N(A) = \{x|x \in G$ and $\forall y \in G, A(xy) = A(yx)\}$. Show that (1) $N(A)$ is a fuzzy subgroup of G and (2) A is a normal fuzzy subgroup of $N(A)$.
19. Let A be a fuzzy subgroup of G and $g \in G$. Define $C(A) \in F(G)$ by: $\forall x \in G, C(A)(x) = A(gxg^{-1})$. Show that (1) $C(A)$ is a fuzzy subgroup of G and (2) A is normal iff $A = C(A)$.
20. Let A be a fuzzy ideal of R. If R has an identity e, show that $A(x) = A(e)$ for any $x \in R$.
21. Prove Proposition 5.27.
22. If A is a fuzzy ideal of R, show that R/A is isomorphic to R/A^*.
23. Prove the assertions in Proposition 5.28.
24. Prove the assertions in Proposition 5.29.
25. Prove the remaining assertions in Proposition 5.30.
26. Show that the class of closed fuzzy sets in a fuzzy topological space satisfies the basic properties C1, C2 and C3.
27. Prove the remaining properties of Lemma 5.2.
28. Prove that closure operator cl in a fuzzy topological space satisfies the basic properties (K1-K4) and vice versa as stated in Propositions 5.33 and 5.34.
29. Provide the proof of Proposition 5.36.
30. Show that the fuzzy Sierpinski space constitutes a fuzzy topological space.
31. Prove that the excluded fuzzy singleton fuzzy topology is indeed a fuzzy topology.
32. Show that the family $\mathcal{T}^*$ does not necessarily define a fuzzy topology.
33. Prove that the included fuzzy set fuzzy topology is indeed a fuzzy topology.
34. Prove that the excluded fuzzy set fuzzy topology is indeed a fuzzy topology.
35. Prove that the Z-fuzzy topology is indeed a fuzzy topology.
36. Show that the following normality properties of the fuzzy Sierpinski space hold:
$$\mathcal{T}_5 \text{ is (weakly) normal} \iff y \in]0, 0.5],$$
$$\mathcal{T}_5 \text{ is (weakly) normal} \iff y \in]0.5, 1],$$
where y is the value of the fuzzy singleton defining $\mathcal{T}_5$.

Chapter 6
Fuzzy Inference and Fuzzy Control

The major subject of this chapter is fuzzy control, one of the most successful application areas of fuzzy set theory. Nowadays, many fuzzy products are visible in the market. Almost every fuzzy product is related to a fuzzy control problem. It is no exaggeration to say that fuzzy set theory is highly accepted partly because of the great success of fuzzy control. Fuzzy controllers may vary substantially dependent on what control problem is to be solved. In this chapter, we just outline the principle of fuzzy control by a typical application case. Since fuzzy inference is the base of fuzzy control, some related concepts such as linguistic variables, hedges, fuzzy propositions, IF-THEN rules and special fuzzy inference approaches are briefly introduced. Considering that fuzzification and defuzzification (especially the latter) play an important role in the design of fuzzy controllers, they also are investigated in this chapter.

6.1 Linguistic Variables and Hedges

A variable is frequently described by words or sentences in a natural or artificial language instead of numerical values in real life. For example, in the statement "John is young", we employ the word "young" rather than a precise number, e.g. 25 or 26, to describe John's age. Similarly, words "very young ", "not young", "old" can also be employed in order to describe "Age". In this sense, "Age " is a variable which can assume the values "young ", "very young ", "not young", "old" etc., which are called linguistic values. Clearly, a linguistic value is a fuzzy set in nature. If a variable can assume linguistic values, then this variable is called a linguistic variable. Besides "Age", "Temperature" is also a linguistic variable which can assume the linguistic values "hot", "cold", "cool", "very hot", "not cold" etc. "The speed of a car" is another example of a linguistic variable since it may take linguistic values "slow", "medium", "fast" etc.

Remark 6.1. *One may use the numerical values 20, 21, 22, 23, 24 etc. to describe a person's age. In this case, the variable "age" which takes numerical* values is regarded as different from the variable "Age" which takes linguistic values. The variable "age" is called the base variable of the variable "Age" by Zadeh [172].

The concept of a linguistic variable plays an important role in representing human imprecise knowledge. The composition of multiple linguistic variables often constitutes a systematic and whole description of some certain knowledge. For example, a person can be regarded as a linguistic variable which is composed of the component linguistic variables "Age", "Height", "Weight", "Appearance", and so on.

The values of a linguistic variable may consist of primary terms such as "young", "old" in the variable "Age" and those which are built from primary ones by using linguistic modifiers or hedges such as "very", "slightly", "more or less", "fairly", "extremely" etc. and the logical connectives "not", "or", "and". A linguistic hedge may be defined as a unary operation on the unit interval. The most commonly used operation h for representing hedges is the α-th power, i.e. $h(a) = a^\alpha$ ($\alpha > 0$). For example, the hedge "very" is often interpreted as the 2-nd power and "more or less" or "fairly" is often interpreted as the 1/2-th power. Now given a unary operation h representing a linguistic hedge and a fuzzy set A on X representing a linguistic value, the linguistic value hA is a fuzzy set on X defined by:

$$\forall x \in X, (hA)(x) = h(A(x)).$$

For example, "very A" and "fairly A" may be respectively defined in the following canonical form:

$$\forall x \in X, \quad (\text{very}A)(x) = A(x)^2$$

$$\forall x \in X, \quad (\text{fairly}A)(x) = \sqrt{A(x)}.$$

Example 6.1. Let $X = \{1, 2, \ldots, 5\}$. The fuzzy set "small" on X is defined by

$$small = 1/1 + 0.8/2 + 0.6/3 + 0.4/4 + 0.2/5.$$

then

$$very\ small = 1/1 + 0.64/2 + 0.36/3 + 0.16/4 + 0.04/5.$$

$$very\ very\ small = 1/1 + 0.4096/2 + 0.1296/3 + 0.0256/4 + 0.0016/5.$$

$$fairly\ small = 1/1 + 0.8944/2 + 0.7746/3 + 0.6325/4 + 0.4472/5.$$

Generally speaking, a hedge operation h is a bijection on $[0, 1]$ such that $h(0) = 0$ and $h(1) = 1$, which indicates that a hedge has no modification to crisp predicates. When $h(a) = a^2$, hA is called the concentration of A. When $h(a) = a^{1/2}$, hA is called the dilatation of A.

6.2 Fuzzy Propositions and IF-THEN Rules

As we know, every proposition has a deterministic truth value either 1 (truth) or 0 (falsity) in mathematical logic. However, if fuzzy concepts are involved in a proposition, then its truth value becomes fuzzy. Consider the proposition "John is young". Since "young" is a fuzzy set, it is often difficult to assign 1 or 0 to express the truth or falsity of the proposition. In this case, it is intuitively reasonable to employ the compatibility degree of John's age with the fuzzy set "young", i.e. the membership degree of John's age in "young" as the truth value of the proposition. If the following definition of "young" is adopted

$$Y(x) = \begin{cases} [1 + (\frac{x-25}{5})^2]^{-1} & \text{if } 25 < x \leq 100 \\ 1 & \text{otherwise} \end{cases},$$

and if John is 30 years old, then the truth value of the proposition "John is young" is $Y(30) = 0.5$. A proposition with its truth value in $[0, 1]$ is called a fuzzy proposition. A classical proposition with the truth value 1 or 0 or a proposition in many-valued logic may be regarded as a special case of fuzzy propositions.

There are two types of fuzzy propositions, atomic propositions and compound propositions. An atomic (fuzzy) proposition is of the form "x is A", where x is a linguistic variable and A is a linguistic value of x. A compound (fuzzy) proposition is a composition of atomic propositions through logical connectives "not", "and", and "or". For example, "John is young" or "John is old" is an atomic proposition, and "John is not young and John is not old" is a compound proposition. In a compound proposition, there may appear multiple linguistic variables. For instance, there are two linguistic variables involved, "Age" and "Height" in the proposition "John is young and John is not tall". Generally, the truth value of a compound proposition can be calculated as follows. Let A and B be linguistic values which are fuzzy sets on X and Y respectively.

(1) The truth value of the proposition "x is not A" is $n(A(x))$, where n is a fuzzy negation.
(2) The truth value of the proposition "x is A and y is B" is $T(A(x), B(y))$, where T is a t-norm.
(3) The truth value of the proposition "x is A or y is B" is $S(A(x), B(y))$, where S is a t-conorm.

The proposition of the form "IF <FP1>, THEN <FP2>" is called a fuzzy IF-THEN rule, where FP1 and FP2 are fuzzy propositions. For instance, "If a person is healthy, then the person is long-lived" is a fuzzy IF-THEN rule. IF-THEN rules are commonly represented by using fuzzy implication operators. In the following we list some operations representing fuzzy IF-THEN rules.

Dienes-Rescher $\quad R_{DR}(x, y) = \max(1 - FP1(x), FP2(y))$.
Lukasiewicz $\quad R_L(x, y) = \min(1, 1 - FP1(x) + FP2(y))$.

Gödel $R_G(x,y) = \begin{cases} 1 & \text{if } FP1(x) \leq FP2(y) \\ FP2(y) & \text{otherwise} \end{cases}$.

Goguen $R_{GO}(x,y) = \begin{cases} 1 & \text{if } FP1(x) \leq FP2(y) \\ \frac{FP2(y)}{FP1(x)} & \text{otherwise} \end{cases}$.

However, some conjunction operations such as $T_{\min}$ and T_π are also employed to represent IF-THEN rules in practice.

Mamdani $R_M(x,y) = \min(FP1(x), FP2(y))$.

Besides, there are other approaches to representing IF-THEN rules among which we mention

Zadeh $R_Z(x,y) = \max\{\min(FP1(x), FP2(y)), 1 - FP1(x)\}$ and

$$R_N(x,y) = \begin{cases} 1 & \text{if } FP1(x) \leq FP2(y) \\ 0 & \text{otherwise} \end{cases}.$$

Hence, an IF-THEN rule is represented by a fuzzy relation.

Example 6.2. Let $X = \{1,2,3,4\}$ and $Y = \{1,2,3\}$. Suppose we know that $x \in X$ is somewhat inversely proportional to $y \in V$. We may use the following IF-THEN rule to express this knowledge.

$$IF\ x\ is\ large, THEN\ y\ is\ small,$$

where the fuzzy sets "large" and "small" are defined as

$$large = 0.1/2 + 0.5/3 + 1/4$$

$$small = 1/1 + 0.5/2 + 0.1/3$$

If the Dienes-Rescher implication is used, then the above IF-THEN rule is represented by

$$R_{DR} = \begin{pmatrix} 1 & 1 & 1 \\ 1 & 0.9 & 0.9 \\ 1 & 0.5 & 0.5 \\ 1 & 0.5 & 0.1 \end{pmatrix},$$

where $R_{DR}(4,3) = 0.1$ may be interpreted as the truth value of the proposition "IF 4 is large, then 3 is small".

Similarly, we have

$$R_L = \begin{pmatrix} 1 & 1 & 1 \\ 1 & 1 & 1 \\ 1 & 1 & 0.6 \\ 1 & 0.5 & 0.1 \end{pmatrix}, \quad R_Z = \begin{pmatrix} 1 & 1 & 1 \\ 0.9 & 0.9 & 0.9 \\ 0.5 & 0.5 & 0.5 \\ 1 & 0.5 & 0.1 \end{pmatrix},$$

$$R_G = \begin{pmatrix} 1 & 1 & 1 \\ 1 & 1 & 1 \\ 1 & 1 & 0.1 \\ 1 & 0.5 & 0.1 \end{pmatrix}, \quad R_M = \begin{pmatrix} 0 & 0 & 0 \\ 0.1 & 0.1 & 0.1 \\ 0.5 & 0.5 & 0.1 \\ 1 & 0.5 & 0.1 \end{pmatrix},$$

$$R_{GO} = \begin{pmatrix} 1 & 1 & 1 \\ 1 & 1 & 1 \\ 1 & 1 & 0.2 \\ 1 & 0.5 & 0.1 \end{pmatrix}, \quad R_N = \begin{pmatrix} 1 & 1 & 1 \\ 1 & 1 & 1 \\ 1 & 1 & 0 \\ 1 & 0 & 0 \end{pmatrix},$$

6.3 Fuzzy Inference Rules

In classical logic, there are three important inference rules called modus ponens, modus tollens, and hypothetical syllogism.

Modus ponens
Given two propositions "x is A" and "IF x is A, THEN y is B", it can be inferred that "y is B". This inference rule is called the modus ponens, which is intuitively expressed in the form of scheme

$$\frac{\begin{array}{c} A \Rightarrow B \\ A \end{array}}{B}$$

Modus tollens
Given two propositions "IF x is A, THEN y is B" and "y is not B", it can be inferred that "x is not A". This inference rule is called the modus tollens, which is intuitively expressed in the form of scheme

$$\frac{\begin{array}{c} A \Rightarrow B \\ \neg B \end{array}}{\neg A}$$

Hypothetical syllogism
Given two propositions "IF x is A, THEN y is B" and "IF y is B, THEN z is C", it can be inferred that "IF x is A, THEN z is C". This inference rule is called the hypothetical syllogism, which is intuitively expressed in the form of scheme

$$\frac{\begin{array}{c} A \Rightarrow B \\ B \Rightarrow C \end{array}}{A \Rightarrow C}$$

In practice, we frequently make imprecise reasoning. For example, from "If a tomato is red, then the tomato is ripe" and "A tomato is slightly red", we may infer that "The tomato is slightly ripe". In this inference, "red" and "slightly red" are not completely matched. In addition, some terms in propositions are frequently fuzzy. The inference process cannot be expressed by means of classical logic. In fuzzy logic, these inference rules are generalized.

Generalized modus ponens
Given two fuzzy propositions "x is A'" and "IF x is A, THEN y is B", it can be inferred that y is B'. This fuzzy inference rule is intuitively expressed in the form of scheme

$$\frac{\begin{array}{c} A \Rightarrow B \\ A' \end{array}}{B'}$$

Generalized modus tollens
Given two fuzzy propositions "IF x is A, THEN y is B" and "y is B'", it can be inferred that x is A'. This fuzzy inference rule is intuitively expressed in the form of scheme

$$\frac{\begin{array}{c} A \Rightarrow B \\ B' \end{array}}{A'}$$

Generalized hypothetical syllogism
Given two fuzzy propositions "IF x is A, THEN y is B" and "IF y is B', THEN z is C", it can be inferred that "IF x is A, THEN z is C'". This fuzzy inference rule is intuitively expressed in the form of scheme

$$\frac{\begin{array}{c} A \Rightarrow B \\ B' \Rightarrow C \end{array}}{A \Rightarrow C'}$$

Remark 6.2. *We adopt a general hypothetical syllogism here. More special hypothetical syllogism: From the propositions "IF x is A, THEN y is B" and "IF y is B, THEN z is C", it is inferred that "IF x is A, THEN z is C'", can be found in [83, 123, 124].*

6.4 The Calculation of Inference Results

In the last section, we introduce three generalized inference rules. Now the question arises: how to calculate the final inference results? More specifically, how to calculate the resulting fuzzy set B' and A' respectively in the generalized modus ponens and the generalized modus tollens, and the resulting relation $A \Rightarrow C'$ in the generalized hypothetical syllogism? Let us start with the generalized modus ponens. Let A and B be fuzzy sets on X and Y respectively. Then the proposition "IF x is A, THEN y is B" is expressed as a

6.4 The Calculation of Inference Results

fuzzy relation R from X to Y. Find the T-composition of A' and R to obtain the conclusion C, i.e.

$$\forall y \in Y, B'(y) = \bigvee_{x \in X} T(A'(x), R(x, y)),$$

where T is a t-norm. In the following, we consider several special cases.

If $T = T_{\min}$ and $R = R_M$, then

$$B'(y) = \bigvee_{x \in X} (A'(x) \wedge R_M(x, y))$$

$$= \bigvee_{x \in X} (A'(x) \wedge A(x) \wedge B(y))$$

$$= (A \odot A') \wedge B(y),$$

where $A \odot A'$ is the inner product of A and A'.

Clearly, this inference approach has the properties:

(1) If $A \cap A' = \emptyset$, then $A \odot A = 0$, and hence $B' = \emptyset$.
(2) If $A = A'$ is normal, then $A \odot A' = 1$, and hence $B' = B$.
(3) If $A' = \{x_0\}$ ($x_0 \in X$), then $A \odot A' = A(x_0)$, and hence $B'(y) = A(x_0) \wedge B(y)$.

The second property means that "x is A" and "IF x is A, THEN y is B" together can infer that "y is B" under the condition that A is normal. This property indicates that the inference approach is the so-called recurring one.

If $T = T_{\min}$ and $R = R_{GO}$, then

$$B'(y) = \bigvee_{x \in X} (A'(x) \wedge R_{GO}(x, y))$$

$$= \bigvee_{A(x) \leq B(y)} A'(x) \vee \bigvee_{A(x) > B(y)} (A'(x) \wedge B(y))$$

$$= \bigvee_{x \in X} A'(x) \wedge ((\bigvee_{A(x) \leq B(y)} A'(x)) \vee B(y))$$

It is easily verified that the approach is recurring.

If $T = T_{\min}$ and $R = R_N$, then

$$B'(y) = \bigvee_{x \in X} (A'(x) \wedge R_N(x, y))$$

$$= \bigvee_{A(x) \leq B(y)} A'(x)$$

In this case, the inference approach has the properties:

(1) If $A' = A$, and hence $B' = B$, which indicates that the approach is a recurring one.

(2) If $A' = $ very A, i.e. $\forall x \in X$, $A'(x) = (A(x))^2$, then $B' = $ very B, i.e. $\forall y \in Y$, $B'(y) = (B(y))^2$.
(3) If $A' = $ more or less A, i.e. $\forall x \in X$, $A'(x) = \sqrt{A(x)}$, then $B' = $ more or less B, i.e. $\forall y \in Y$, $B'(y) = \sqrt{B(y)}$.

The proof of these properties is straightforward and is omitted here.

For the generalized modus tollens, the mathematical representation of reasoning is similar. Firstly, "IF x is A, THEN y is B" is expressed by a fuzzy relation R. Then combine B' and R by T-composition to obtain the conclusion A':

$$A'(x) = \bigvee_{y \in Y} B'(y) \wedge R(x, y).$$

If $T = T_{\min}$ and $R = R_M$, then

$$A'(x) = \bigvee_{y \in Y} B'(y) \wedge A(x) \wedge B(y)$$
$$= (\bigvee_{y \in Y} (B(y) \wedge B'(y))) \wedge A(x)$$
$$= (B \odot B') \wedge A(x).$$

If $T = T_{\min}$ and $R = R_N$, then

$$A'(x) = \bigvee_{y \in Y} (B'(y) \wedge R_N(x, y))$$
$$= \bigvee_{A(x) \leq B(y)} B'(y)$$

When $B' = B^c$,

$$A'(x) = \bigvee_{A(x) \leq B(y)} B^c(y) = \bigvee_{A^c(x) \geq B^c(y)} B^c(y) = A^c(x).$$

This property means that from "IF x is A, THEN y is B" and "y is not B" we can infer that "x is not A", which coincides with classical logic.

Similarly, it can be easily verified that

(1) From "IF x is A, THEN y is B" and "y is not very B" we can infer that "x is not very A".
(2) From "IF x is A, THEN y is B" and "y is not more or less B" we can infer that "x is not more or less A".

Now, let us turn to the generalized hypothetical syllogism.

Firstly, "IF x is A, THEN y is B" and "IF y is B', THEN z is C" are respectively expressed by R_1 and R_2. Then the fuzzy proposition R: "IF x is A, THEN z is C'" is represented by the round composition of R_1 and R_2, i.e. $R = R_1 \circ R_2$.

6.4 The Calculation of Inference Results

For example, if $R_1 = R_2 = R_M$, then

$$R(x,z) = \bigvee_{y \in Y} (R_1(x,y) \wedge R_2(y,z))$$

$$= \bigvee_{y \in y} (A(x) \wedge B(y) \wedge B'(y) \wedge C(z))$$

$$= \bigvee_{y \in y} (B(y) \wedge B'(y)) \wedge (A(x) \wedge C(z))$$

$$= (B \odot B') \wedge (A(x) \wedge C(z)).$$

If $B = B'$ is normal, then $B \odot B' = 1$ and thus $R(x,z) = A(x) \wedge C(z) = R_M(x,z)$, which means that from "IF x is A, THEN y is B" and "IF y is B, THEN z is C" we can infer that "IF x is A, THEN z is C".

In the generalized modus ponens and tollens, there may be more than one IF-THEN rule. For example, the generalized modus ponens may be of the form:

$$A_1 \Rightarrow B_1$$
$$A_2 \Rightarrow B_2$$
$$\ldots\ldots$$
$$A_n \Rightarrow B_n$$
$$\frac{A'}{B'}$$

Suppose that $A_i \Rightarrow B_i$ $(i = 1, 2, \ldots, n)$ is represented by R_i $(i = 1, 2, \ldots, n)$. Then the max operation is employed to combine $R_1, R_2, \ldots, R_n$ to form a single fuzzy relation R, i.e. $R(x,y) = \max(R_1(x,y), R_2(x,y), \ldots, R_n(x,y))$. The conclusion B' in the generalized modus ponens is calculated as

$$B'(y) = \bigvee_{x \in X} (T(A'(x), R(x,y))).$$

If $T = T_{\min}$ and $R_i = R_M$ $(i = 1, 2, \ldots, n)$, then

$$B'(y) = \bigvee_{x \in X} (A'(x) \wedge R(x,y))$$

$$= \bigvee_{i=1}^{n} \bigvee_{x \in X} (A_i(x) \wedge A'(x)) \wedge B_i(y)$$

$$= \bigvee_{i=1}^{n} ((A_i \odot A') \wedge B_i(y))$$

The conclusion A' in the generalized modus tollens is similarly calculated as

$$A'(x) = \bigvee_{y \in Y} (B'(y) \wedge R(x,y))$$

$$= \bigvee_{i=1}^{n} \bigvee_{y \in Y} (B_i(y) \wedge B'(y)) \wedge A_i(x)$$

$$= \bigvee_{i=1}^{n} ((B_i \odot B') \wedge A_i(x))$$

Remark 6.3. *There exist some debates over which operation should be used to aggregate $R_1, R_2, \cdots, R_n$ into R. The choice of max is based on the understanding: all IF-THEN rules are disjunctive, i.e. $A_1 \Rightarrow B_1$ or $A_2 \Rightarrow B_2$ or $\cdots$ or $A_n \Rightarrow B_n$.*

For the generalized hypothetical syllogism, a more general form may be as follows:

$$A_1 \Rightarrow A_2$$
$$A_2' \Rightarrow A_3$$
$$\cdots\cdots$$
$$\underline{A_{n-1}' \Rightarrow A_n}$$
$$A_1 \Rightarrow A_n'$$

Let A_i and A_i' be fuzzy sets on X_i ($i = 2, \cdots, n$) and let A_1 be a fuzzy set on X_1. If $A_1 \Rightarrow A_2$, $A_2' \Rightarrow A_3$, $\cdots$, $A_{n-1}' \Rightarrow A_n$ and $A_1 \Rightarrow A_n'$ is respectively represented by $R_1, R_2, \cdots, R_n$ and R, then R can be calculated as $R = (((R_1 \circ_T R_2) \circ_T R_3) \cdots \circ_T R_n)$. More specifically, let $T = T_{\min}$ and $R_i = R_M$. Then R is calculated as

$$R(x_1, x_n) = \bigvee_{\substack{x_i \in X_i, \\ i=2,3,\cdots,n-1}} (A_1(x_1) \wedge A_2(x_2) \wedge A_2'(x_2) \wedge \cdots \wedge A_{n-1}'(x_{n-1}) \wedge A_n(x_n))$$

$$= A_1(x_1) \wedge A_n(x_n) \wedge (A_2 \odot A_2') \wedge \cdots \wedge (A_{n-1} \odot A_{n-1}')$$

6.5 Fuzzification and Defuzzification

A fuzzification function, f, is a function that transforms every real number within the range of a given variable into a fuzzy quantity (mainly fuzzy number) that approximates the real number. That is, for each value x^*, $f(\cdot, x^*)$ is a function serving as a fuzzy approximation of x^*. The following are two extensively used fuzzification approaches:

6.5 Fuzzification and Defuzzification

Gaussian fuzzification

In this method, the measurement x^* is mapped into the function of the form

$$A(x) = e^{-\left(\frac{x-x^*}{a}\right)^2},$$

where $a > 0$ is a real number.

Triangular fuzzification

In this method, the measurement x^* is mapped into the triangular fuzzy number $A = (x^* - b, x^*, x^* + b)$, where $b > 0$ is a real number.

The motivation behind fuzzification is that it may be more realistic to replace a crisp measurement x^* by a function which means "around x^*" in considering the measurement uncertainty due to the degrading sensors, environment disturbances or other reasons. In some fuzzy controllers, input variables are not fuzzified and their measurements are directly employed as facts. In this case, it is equivalent to the following fuzzification:

$$A(x) = \begin{cases} 1 & x = x^* \\ 0 & \text{otherwise} \end{cases}.$$

As we know in the last section, the results in fuzzy inferences are fuzzy sets which present an imprecise description of the system. These fuzzy sets must be converted into crisp numbers so that definite actions can be taken in a control system. The defuzzification is to reduce a fuzzy set to a number which in some sense summarizes the fuzzy set. Among the various defuzzification methods existing in the literature, the most frequently used one is the center of gravity method(COG) or centroid method. Let A be a fuzzy set on X (X is a bounded subset of $\mathbb{R}$ in most applications). In COG, the centroid of the area of A is used to represent A. Hence the defuzzification is expressed by the formula

$$COG(A) = \frac{\int_X xA(x)dx}{\int_X A(x)dx}.$$

Sometimes, to drop smaller membership degrees of A, $COG(A)$ is modified as

$$COG_\alpha(A) = \frac{\int_{A_\alpha} xA(x)dx}{\int_{A_\alpha} A(x)dx}.$$

Example 6.3. Let the fuzzy set A on $X = [0, 1]$ be defined by

$$A(x) = \begin{cases} 5x & 0 \leq x \leq 0.2 \\ \frac{5}{4}(1-x) & 0.2 < x \leq 1 \end{cases}$$

For $0 < \alpha < 1$,

$$A_\alpha = \left[\frac{\alpha}{5}, 1 - \frac{4}{5}\alpha\right].$$

Hence

$$\int_{A_\alpha} A(x)dx = 5\int_{0.2\alpha}^{0.2} xdx + \frac{5}{4}\int_{0.2}^{1-0.8\alpha}(1-x)dx$$

$$= \frac{5}{2}x^2\Big|_{0.2\alpha}^{0.2} + \frac{5}{4}\left(x - \frac{x^2}{2}\right)\Big|_{0.2}^{1-0.8\alpha}$$

$$= 0.5(1-\alpha^2)$$

Similarly,

$$\int_{A_\alpha} xA(x)dx = 5\int_{0.2\alpha}^{0.2} x^2 dx + \frac{5}{4}\int_{0.2}^{1-0.8\alpha} x(1-x)dx$$

$$= 0.2(1 - 2\alpha^2 + \alpha^3)$$

Therefore,

$$COG_\alpha(A) = \frac{2(1 - 2\alpha^2 + \alpha^3)}{5(1-\alpha^2)} \text{ and } COG(A) = COG_0(A) = 0.4.$$

Besides COG, there exist other defuzzification methods. For example, the methods FOM (first of maxima) and LOM (last of maxima) are possible alternatives. which employ $FOM(A) = \inf\{x|x \in \text{core}(A)\}$ and $LOM(A) = \sup\{x|x \in \text{core}(A)\}$ respectively to represent A, where $\text{core}(A) = \{x|A(x) = \sup_{x \in X} A(x)\}$.

To evaluate a defuzzification operator D, the following criteria are suggested [143]. Let A and B be fuzzy sets on a subset X of $\mathbb{R}$.

(1) Core selection: $D(A) \in \text{core}(A)$.
(2) Scale invariance: This criterion means that the defuzzification operator must be invariant under the permissible scale transformation. For example,
$$D(a \cdot A + b) = D(A), (a \in]0, +\infty[, b \in \mathbb{R}),$$
where $a \cdot A + b$ is defined by $\forall x \in X, (a \cdot A + b)(x) = a \cdot A(x) + b$, which is called the interval scale invariance.

In the case $b = 0$, $D(a \cdot A) = D(A)$, which is called the ratio scale invariance.

In the case $a = 1$, $D(A+b) = D(A)$, which is called the relative scale invariance.
(3) Monotonicity: If $B(D(A)) = A(D(A))$ and $\forall x < D(A), B(x) \leq A(x)$ and $\forall x > D(A), B(x) \geq A(x)$, then $D(B) \geq D(A)$.
(4) t-conorm criterion: If $D(A) \leq D(B)$, then $D(A) \leq D(A \cup_S B)D(B)$, where S is a t-conorm.

In addition, continuity (robustness), computational efficiency etc. are also considered in evaluating a defuzzification method. Using these criteria, we

6.6 The Principle of Fuzzy Control

can evaluate COG, FOM and LOM. COG does not fulfill any of criteria (1)-(4) except monotonicity. Both FOM and LOM satisfy all of (1)-(4). We leave the proofs of these assertions as exercises. It should be pointed out that which defuzzification method is the best depends heavily on the type of applications.

6.6 The Principle of Fuzzy Control

As we know, classical controllers are useful in the industry, engineering and space science etc. However, precise mathematical models need to be constructed to implement a control process. For a complex time-varying and non-linear system, it is often difficult or even impossible to acquire such mathematical models and necessary precise data. On the contrary, an experienced operator may work well when faced with such a system. Instead of precise models, fuzzy controllers are based on fuzzy inference rules elicited from human experience.

Typically, a process of fuzzy control may involve four main steps. In the first step, the measurements of all variables are taken, which normally are crisp numbers and their linguistic values are determined. In the second step, the measurements are transformed into fuzzy numbers, which is the fuzzification process. This step may be omitted if the measurements are reliable and thus can be directly employed. In the third step, the fuzzy control rules are established to form a fuzzy rule base and approximate reasonings are made using the approaches in the last section. The evaluation results are generally fuzzy sets which are defuzzified in the final step. The resulting values after defuzzification represent actions taken by the fuzzy controller. In the following, we illustrate these steps by a simple problem, the control of inverted pendulum, which often serves as a typical case to illustrate the merits of a new method in the area of both classical and fuzzy control.

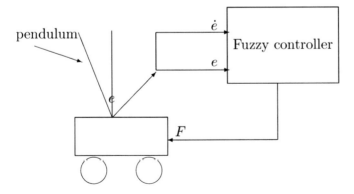

Fig. 6.1 An inverted pendulum

The inverted pendulum means a movable pole attached to a vehicle through a pivot as depicted in Fig. 6.1. The control problem is to keep the pole (pendulum) vertically by moving the vehicle to the left or right in an appropriate velocity.

In this problem, two input variables are involved. One is the angle (error) between the actual position of the pole and the vertical position, denoted e, which is measured by an angle sensor. The other is the rate of change of e, denoted $\dot{e}$, which is calculated by the use of two successive values of e. The output variable F is the force applied to the vehicle. When the pole is tilted to the left (right) of the vertical position, e is defined as negative (positive), and a similar convention applies to $\dot{e}$. The force F is defined as negative (or positive) when it is towards left (right).

Firstly, choose a linguistic description of the states of the input variables. In this example, the following canonical linguistic states may be employed:

NL – negative large

NM – negative medium

NS – negative small

AZ – approximately zero

PL – positive large

PM – positive medium

PS – positive small.

For example, e =NL means that the pole is tilted to the left with a large angle; F =NL means a large force towards left applied to the pole etc. These linguistic values may be represented by the triangular fuzzy numbers shown in Fig. 6.2.

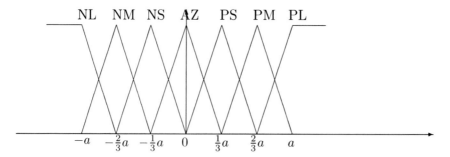

Fig. 6.2 Fuzzy numbers representing linguistic values of variables

Remark 6.4. *In Fig.6.2, $[-a, a]$ stands for the range of the variables. For example, the range of e is between $-\pi$ and π, and thus $a = \pi$ for e.*

6.6 The Principle of Fuzzy Control

Secondly, a fuzzification function is selected for each input variable to fuzzify the input value e and $\dot{e}$. We assume that e and $\dot{e}$ are respectively fuzzified as $f_e(x)$ and $f_{\dot{e}}(y)$.

Thirdly, formulate the fuzzy control rules according to human experience. In our example, all control rules are of the canonical form:

$$\text{If } e = A \text{ and } \dot{e} = B, \text{ then } F = C,$$

where A, B, C represent linguistic states. Since there are two input variables and each input variable has seven linguistic states, 49 possible fuzzy control rules are available, which are organized in the following table.

Table 6.1 Fuzzy control rules for stabilizing an inverted pendulum

F	NL	NM	NS	AZ	PS	PM	PL
NL	NL	NL	NL	NL	NM	NS	AZ
NM	NL	NL	NM	NM	NS	AZ	PS
NS	NL	NM	NS	NS	AZ	PS	PM
AZ	NL	NM	NS	AZ	PS	PM	PL
PS	NM	NS	AZ	PS	PS	PM	PL
PM	NS	AZ	PS	PM	PM	PL	PL
PL	AZ	PS	PM	PL	PL	PL	PL

In the table, the first row displays the values taken by $\dot{e}$ except the first entry and the first column displays the values taken by e except the first entry. The others are the values of F. All the control rules are established based on the intuition. For example, if the pole tilts towards left for a small angle (e =NS) and its rate of change is decreasingly small ($\dot{e}$=NS), then the vehicle should move towards left in a low velocity, i.e. F =NS. The other rules can be similarly explained intuitively.

Fourthly, fuzzified measurements serve as the input of the fuzzy controller. An approximate reasoning approach is selected to make inferences in order to obtain the output. If the Mamdani's approach is adopted, i.e. "If $e = A_i$ and $\dot{e} = B_i$, then $F = C_i$ $(i = 1, 2, \ldots, n)$" are converted into

$$R(x, y, z) = \bigvee_{i=1}^{n} (A_i(x) \wedge B_i(y) \wedge C_i(z)),$$

then the output C is computed as

$$C(z) = \bigvee_{x \in X, y \in Y} (R(x, y, z) \wedge f_e(x) \wedge f_{\dot{e}}(y))$$

$$= \bigvee_{x \in X, y \in Y} \bigvee_{i=1}^{n} (A_i(x) \wedge B_i(y) \wedge C_i(z) \wedge f_e(x) \wedge f_{\dot{e}}(y))$$

$$= \bigvee_{i=1}^{n} \bigvee_{x \in X, y \in Y} (A_i(x) \wedge B_i(y) \wedge C_i(z) \wedge f_e(x) \wedge f_{\dot{e}}(y))$$

$$= \bigvee_{i=1}^{n} (((A_i \odot f_e) \wedge (B_i \odot f_{\dot{e}})) \wedge C_i(z)),$$

where X and Y are respectively the range of x and y.

Finally, a defuzzification approach is selected to transform the resulting fuzzy set C to a single number, which determines the action taken by the fuzzy controller.

Although practical control problems may be various, the basic principle of fuzzy control remains the same. Of course, there may exist multiple alternatives in each step of fuzzy control, e.g. linguistic states in Step 1, the number of control rules in Step 2, fuzzification approaches in step 3, fuzzy inference approaches in Step 4 and finally defuzzification approaches. Which approaches are chosen should depend on the concrete control problem.

Remark 6.5. *The concept of a linguistic variable was proposed by Zadeh in 1975 [173]. The compositional rule of inference can be found in [174]. For a comparative study of different fuzzy inference methods, see [44, 102, 152]. An exhaustive overview of approximate reasoning by using fuzzy sets can be found in [47, 48], which collected all research before 1991. For the extensive overview and mutual comparisons between different defuzzification approaches, the reader may refer to [143]. In 1975, Mamdani and Assilian designed the first fuzzy controller to control a steam engine [101]. In 1978, Holmblad and Østergaard [68] developed a fuzzy cement kiln controller which was the first fuzzy controller for the control of a industrial process. Many successful applications of fuzzy control were implemented in Japan [67, 139, 166, 167, 168]. In [138, 141], some applications of fuzzy control are collected.*

6.7 Exercises

1. Let $X = \{x_1, x_2, x_3\}$ and $Y = \{y_1, y_2\}$. $A = 0.5/x_1 + 0.8/x_2 + 1/x_3$ and $B = 1/y_1 + 0.4/y_2$. Use $R_{DR}, R_L, R_D, R_Z, R_{GO}, R_N$ and R_M respectively to represent the proposition "IF x is A, THEN y is B".
2. Given $T = \min$ and $R = R_Z$, find B' for A' =very A and A' =more or less A respectively in the generalized modus ponens.
3. Given $A(x) = \dfrac{x+1}{3}$, $B(y) = 1 - y$, $A'(x) = 1 - x$ ($x, y \in [0, 1]$), calculate B' in the generalized modus ponens if $T = \min$ and $R = R_M$.
4. Given $T = \min$, $A' = A$, $R = R_L$, find B' in the generalized modus ponens.
5. Let $T = T_L$, $R = R_L$ and $B' = B^c$. If $\text{plt}(B) = 0$, show that $A' = A^c$ in the generalized modus tollens.

6.7 Exercises

6. Let $T = T_{\min}$ and $R = R_{DR}$. Find A' in the generalized modus tollens for $B' = $ not B, $B' = $ not very B and $B' = $ not more or less B respectively.
7. If $R_1 = R_2 = R_L$, find R in the generalized hypothetical syllogism.
8. If $R_1 = R_2 = R_N$, find R in the generalized hypothetical syllogism. Can we infer that "IF x is A, THEN z is C" from "IF x is A, THEN y is B" and "IF y is B, THEN z is C" in this case?
9. Give formulae to calculate C and C' in the following inference schemes:

$$\frac{(A_1 \text{ and } A_2) \Rightarrow B}{C} \qquad \frac{(A_1 \text{ and } A_2) \Rightarrow B}{C'}$$

10. Show that COG does not fulfil core selection, scale invariance and t-conorm criterion.
11. Show that both FOM and LOM satisfy core selection, scale invariance, t-conorm criterion and monotonicity.
12. Consider the following fuzzy inference rules:

    IF x is A_1, THEN y is A_2
    IF x is A_2, THEN y is A_1,

    where $A_1 = (-1, 0, 1)$ and $A_2 = (0, 1, 2)$ are triangular fuzzy numbers.

    (1) Let $T = T_{\min}$, $R_1 = R_2 = R_M$. If the input is $x^* = 0.3$, find the conclusion C.
    (2) Fuzzify the input $x^* = 0.3$ using the triangular fuzzification approach by choosing a proper parameter b and find C.
    (3) Defuzzify C in (2) using COG and FOM respectively.

13. Design a fuzzy controller to maintain the height of water in a container with an outlet and an inlet at a desired level.

References

[1] Abd El-Monsef, M.E., Ghanim, M.H., Kerre, E.E., Mashhour, A.S.: Fuzzy topological results. In: Proceedings of the 5th Prague Topological Symposium, pp. 1–5. Heldermann Verlag, Berlin (1983)
[2] Adamo, J.M.: Fuzzy decision trees. Fuzzy Sets and Systems 4, 207–219 (1980)
[3] Akgul, M.: Some properties of fuzzy groups. J. Math. Anal. Appl. 133, 93–100 (1988)
[4] Alsina, C., Trillas, E.: When (S, N)-implication are (T, T_1)-conditional functions? Fuzzy Sets and Systems 134, 305–310 (2003)
[5] Anthony, J.M., Sherwood, H.: Fuzzy groups redefined. J. Math. Anal. Appl. 69, 124–130 (1979)
[6] Baas, S.M., Kwakernaak, H.: Rating and ranking of multiple-aspect alternatives using fuzzy sets. Automatic 13, 47–58 (1977)
[7] Baczynski, M., Jayaram, B.: Residual implications revisted. Notes on the Smets-Magrez Theorem. Fuzzy Sets and Systems 145, 267–277 (2004)
[8] Baczynski, M., Jayaram, B.: On the characterizations of (S, N)-implications. Fuzzy Sets and Systems 158, 1713–1727 (2007)
[9] Baczynski, M., Jayaram, B.: (S, N)- and R-implications: A state-of-the-art survey. Fuzzy Sets and Systems 159, 1836–1859 (2008)
[10] Baldwin, J.F., Guild, N.C.F.: Comparison of fuzzy sets on the same decision space. Fuzzy Sets and Systems 2, 213–231 (1979)
[11] Bandler, W., Kohout, L.: Semantics of implication operators and fuzzy relational products. Int. J. Man-Mach. Studies 12, 89–116 (1980)
[12] Bandler, W., Kohout, L.: Fuzzy power sets and fuzzy implication operators. Fuzzy Sets and Systems 4, 13–30 (1980)
[13] Bandler, W., Kohout, L.: Special properties, Closures and interiors of crisp and fuzzy relations. Fuzzy Sets and Systems 26, 317–331 (1988)
[14] Banerjee, A.: Rational choice under fuzzy preference: the Orlovsky choice function. Fuzzy Sets and Systems 53, 295–299 (1993)
[15] Bhakat, S.K., Das, P.: Fuzzy subrings and ideals redefined. Fuzzy Sets and Systems 81, 383–393 (1996)
[16] Bortolan, G., Degni, R.: A review of some methods for ranking fuzzy subsets. Fuzzy Sets and Systems 15, 1–19 (1985)

[17] Bourke, M., Fisher, D.G.: Solution algorithms for fuzzy relational equations with max-product composition. Fuzzy Sets and Systems 94, 61–69 (1998)
[18] Bouyssou, D.: Acyclic fuzzy preference and the Orlovsky choice function. Fuzzy Sets and Systems 89, 107–111 (1997)
[19] Buckley, J.J.: Fuzzy hierrarchical analysis. Fuzzy Sets and Systems 17, 233–247 (1985)
[20] Bufardi, A.: On the construction of fuzzy preference structures. Journal of Multi-criteria Decision and Analysis 7, 169–175 (1998)
[21] Bufardi, A.: On the fuzzification of the classical definition of preference structure. Fuzzy Sets and Systems 104, 323–332 (1999)
[22] Campos, L., Munoz, A.: A subjective approach for ranking fuzzy numbers. Fuzzy Sets and Systems 29, 145–153 (1989)
[23] Campos, L., Gonzalez, A., Vila, M.A.: On the use of the ranking function approach to solve fuzzy matrix games in a direct way. Fuzzy Sets and Systems 49, 193–203 (1992)
[24] Campos, L., Verdegay, J.L.: Linear programming problems and ranking of fuzzy numbers. Fuzzy Sets and Systems 32, 1–11 (1989)
[25] Chang, C.L.: Fuzzy topological spaces. J. Math. Anal. Appl. 24, 182–190 (1968)
[26] Chang, W.: Ranking of fuzzy utilities with triangular membership functions. In: Proceedings of International Conference on Policy Analysis and Systems, pp. 263–272 (1981)
[27] Chen, S.: Ranking fuzzy numbers with maximizing set and minimizing set. Fuzzy Sets and Systems 17, 113–129 (1985)
[28] Chen, S., Hwang, C.: Fuzzy Multiple Attribute Decision Making. Springer, Heidelberg (1992)
[29] Czogala, E., Pedrycz, W.: Controll problems in fuzzy systems. Fuzzy Sets and Systems 7, 257–273 (1982)
[30] Das, P.S.: Fuzzy groups and level subgroups. J. Math. Anal. Appl. 84, 264–269 (1981)
[31] De Baets, B.: Idempotent uninorms. Europ. J. Oper. Research 118, 631–642 (1999)
[32] De Baets, B.: An order-theoretic approach to solving $\sup -\mathcal{T}$ equations. In: Ruan, D. (ed.) Fuzzy Set Theory and Advanced Mahematical Applications, pp. 67–87. Kluwer Academic Publishers, Dordrecht (1995)
[33] De Baets, B.: Analytical solution methods for fuzzy relation equations. In: Dubois, D., Prade, H. (eds.) The Handbooks of Fuzzy Sets Series, vol. 1, pp. 291–340. Kluwer Academic Publishers, Dordrecht (2000)
[34] De Baets, B., Van de Walle, B.: Weak and strong fuzzy interval orders. Fuzzy Sets and Systems 79, 213–225 (1996)
[35] De Baets, B., Van de Walle, B., Kerre, E.: Fuzzy preference structures without incomparability. Fuzzy Sets and Systems 76, 333–348 (1995)
[36] Delgado, M., Verdegay, J.L., Vila, M.A.: A general mode for linear programming. Fuzzy Sets and Systems 29, 21–29 (1989)
[37] Delgado, M., Verdegay, J.L., Vila, M.A.: A procedure for ranking fuzzy numbers. Fuzzy Sets and Systems 26, 49–62 (1988)
[38] Dib, K.A., Galham, N., Hassam, A.M.: Fuzzy rings and fuzzy ideals. J. Fuzzy Math. 4, 245–261 (1996)

References

[39] Di Nola, A.: On solving realtional equations in Brouwerian lattices. Fuzzy Sets and Systems 34, 365–376 (1990)

[40] Di Nola, A., Sessa, S., Pedrycz, W., Sanchez, E.: Fuzzy Relation Equations and Their Applications to Knowledge Engineering. In: System Theory, Knowledge Engineering and Problem Solving. Kluwer Academic Publishers, Dordrecht (1989)

[41] Di Nola, A.: Fuzzy equations in infinitely distributive lattices. In: Proc. Second IFSA World Congress, Tokyo, pp. 533–534 (1987)

[42] Di Nola, A., Pedrycz, W., Sessa, S., Wang, P.Z.: Fuzzy relation equations as a basis of fuzzy modelling: an overview. Fuzzy Sets and Systems 40, 415–429 (1990)

[43] Dixit, V.N., Kumar, R., Ajmal, N.: On fuzzy rings. Fuzzy Sets and Systems 49, 205–213 (1992)

[44] Driankov, D., Hellendoorn, H., Reinfrank, M.: An introduction to Fuzzy Control. Springer, Berlin (1993)

[45] Dubois, D., Prade, H.: Fuzzy Sets and Systems: Theory and Applications. Academic Press, New York (1980)

[46] Dubois, D., Prade, H.: Ranking fuzzy numbers in the setting of possibility theory. Information Sciences 30, 183–224 (1983)

[47] Dubois, D., Prade, H.: Fuzzy sets in approximate reasoning, Part 1: Inference with possibility distributions. Fuzzy Sets and Systems 40, 143–201 (1991)

[48] Dubois, D., Prade, H.: Fuzzy sets in approximate reasoning, Part 2: Logic approaches. Fuzzy Sets and Systems 40, 202–244 (1991)

[49] Dubois, D., Prade, H.: The use of fuzzy numbers in decision analysis. In: Gupta, M.M., Sanchez, E. (eds.) Fuzzy Information and Decision Processes, pp. 309–321. North-Holland Publishing Company, Amsterdam (1982)

[50] Dubois, D., Prade, H.: Operations on fuzzy numbers. International Journal of System Science 9, 613–626 (1978)

[51] Dubois, D., Prade, H.: Fuzzy numbers: an overview. In: Dubois, D., Prade, H. (eds.) Readings in Fuzzy Sets for Intelligent Systems, pp. 112–148. Morgan Kaufmann Publishers, San Francisco (1993)

[52] Dubois, D., Prade, H.: Fuzzy real algebra: some results. Fuzzy Sets and Systems 2, 327–348 (1979)

[53] Dubois, D., Prade, H.: A review of fuzzy set aggregation connectives. Information Science 36, 85–121 (1985)

[54] Fodor, J.: An axiomatic approach to fuzzy preference modeling. Fuzzy Sets and Systems 52, 47–52 (1992)

[55] Fodor, J., Yager, R., Rybalov, A.: Structure of Uninorms. Int. J. of Uncertainty, Fuzziness and Knowledge-based Systems 5, 411–427 (1997)

[56] Fodor, J., Roubens, M.: Fuzzy preference modelling and multicriteria decision support. Kluwer Academic Publishers, Dordrecht (1994)

[57] Fodor, J.: On fuzzy implication operators. Fuzzy Sets and Systems 42, 293–300 (1991)

[58] Fodor, J.: A new look at fuzzy connectives. Fuzzy Sets and Systems 57, 141–148 (1993)

[59] Fortemps, P., Roubens, M.: Ranking and defuzzification methods based on area compensation. Fuzzy Sets and Systems 82, 319–330 (1996)

[60] Freeling, S.: Fuzzy sets and decision analysis. IEEE Transactions on Systems, Man and Cybernetics 10, 341–354 (1980)

[61] Ghanim, M.H., Mashhour, A.S.: Characterization of fuzzy topologies by s-neighborhood systems. Fuzzy Sets and Systems 9, 211–214 (1983)
[62] Goguen, J.A.: L-fuzzy sets. J. Math. Anal. Appl. 8, 145–174 (1967)
[63] Gottwald, S.: Approximately solving fuzzy relational equations: some mathematical results and some heuristic proposals. Fuzzy Sets and Systems 66, 175–193 (1994)
[64] Gu, J., Wang, X.: A class of methods for ranking alternatives. Journal of Systems Engineering 3, 14–23 (1988) (in Chinese)
[65] Gupta, M.M., Qi, J.: Design of fuzzy fuzzy logic controllers based on generalized T-operators. Fuzzy Sets and Systems 40, 473–489 (1991)
[66] Ha, M., Wu, C.: Theory of Fuzzy Measure and Fuzzy integral. Science Press (1998) (in Chinese)
[67] Hirota, K., Arai, A., Hachisu, S.: Fuzzy controlled robot arm playing two-dimensional ping-pong game. Fuzzy Sets and Systems 32, 149–159 (1989)
[68] Holmblad, L.P., Østergaard, J.J.: Control of a cement kiln by fuzzy logic. In: Gupta, M., Sanchez, E. (eds.) Fuzzy Information and Decision Processes, pp. 398–409. North-Holland Publishing Company, Amsterdam (1982)
[69] Hutton, B.: Normality in fuzzy topological spaces. J. Math. Anal. Appl. 50, 74–79 (1975)
[70] Jain, R.: Decision making in the presence of fuzzy variables. IEEE Trans. on Systems, Man, and Cybernetics 6, 698–703 (1976)
[71] Jenei, S., Fodor, J.C.: On continous triangular norms. Fuzzy Sets and Systems 100, 273–282 (1998)
[72] Kim, K., Park, K.S.: Ranking fuzzy numbers with index of optimism. Fuzzy Sets and Systems 35, 143–150 (1990)
[73] Kerre, E.: Introduction to the Basic Principles of Fuzzy Set Theory and Some of Its applications. Communication & Cognition, Gent (1993)
[74] Kerre, E.: Fuzzy Sets and Approximate Reasoning. Xian Jiaotong University Press, Xian (1999)
[75] Kerre, E.: The use of fuzzy set theory in electrocardiological diagnostics. In: Gupta, M., Sanchez, E. (eds.) Approximate reasoning in decision-analysis, pp. 277–282. North-Holland Publishing Company, Amsterdam (1982)
[76] Kerre, E.: Characterization of normality in fuzzy topological spaces. Simon Stevin 53, 239–248 (1979)
[77] Kerre, E., Ottoy, P.L.: Lattice properties of neighborhood systems in a Chang fuzzy topological space. Fuzzy Sets and Systems 30, 205–213 (1989)
[78] Kerre, E., Ottoy, P.L.: On α-generated fuzzy topologies. Fasciculi Mathematica 19, 127–134 (1990)
[79] Kerre, E., Ottoy, P.L.: On the different notions of neighbourhood in Chang-Goguen fuzzy topological spaces. Simon Stevin 61, 131–146 (1987)
[80] Klement, E.P., Mesiar, R., Pap, E.: A characterization of the ordering of continuous t-norms. Fuzzy Sets and Systems 86, 189–195 (1997)
[81] Klement, E.P., Mesiar, R., Pap, E.: Triangular Norms, Trends in Logic. Kluwer Academic Publishers, Dordrecht (2000)
[82] Klir, G.J.: Fuzzy Sets: An Overview of Fundamentals, Applications and Personal Views. Beijing Normal University Press, Beijing (2000)
[83] Klir, G.J., Yuan, B.: Fuzzy Sets and Fuzzy Logic – Theory and Applications. Prentice Hall PTR, New Jersey (1995)

[84] Kohout, L.J., Bandler, W.: Fuzzy relational product as a tool for analysis and synthesis of the behaviour of complex natural and artificial systems. In: Fuzzy Sets: Theory and Applications to Policy Analysis and Information Systems, pp. 341–367. Plenum Press, New York (1980)

[85] Kolodziejczyk, W.: Orlovsky's concept of decision-making with fuzzy preference relation – further results. Fuzzy Sets and Systems 19, 197–212 (1990)

[86] Kumar, R.: Fuzzy subgroups, fuzzy ideals, and fuzzy cosets: some properties. Fuzzy Sets and Systems 48, 267–274 (1992)

[87] Laarhoven, P.J.M., Pedrycz, W.: A fuzzy extension of Saaty's priority theory. Fuzzy Sets and Systems 11, 229–241 (1983)

[88] Ling, C.H.: Representation of associative functions. Publ. Math. Debrecen 12, 189–212 (1965)

[89] Liou, T., Wang, J.: Ranking fuzzy numbers with integral value. Fuzzy Sets and Systems 50, 247–255 (1992)

[90] Liu, P., Wu, M.: Fuzzy Theory and Its Applications. National University of Defence Technology Press, Changsha (1998) (in Chinese)

[91] Liu, W.: Fuzzy invariant subgroups and fuzzy ideals. Fuzzy Sets and Systems 8, 133–139 (1982)

[92] Liu, Y.M., Luo, M.K.: Fuzzy topology, Advances in Fuzzy Systems – Applications and Theory. World Scientific, Singapore (1997)

[93] Llamazares, B.: Characterization of fuzzy preference structures through Lukasiewicz triplets. Fuzzy Sets and Systems 136, 217–235 (2003)

[94] Lowen, R.: Fuzzy topological spaces and fuzzy compactness. J. Math. Anal. Appl. 56, 621–633 (1976)

[95] Lowen, R.: On the existence of natural non-topological fuzzy topological spaces. Heldermann Verlag, Berlin (1985)

[96] Lu, J., Zhang, G., Ruan, D., Wu, F.: Multi-objective Group Decision Making. Imperial College Press, Singapore (2007)

[97] Ludescher, H., Róventa, E.: Sur les topologies floues definie à l'aide des voisinages, pp. 575–577. C.R. Acad. Sc., Paris (1976)

[98] Luo, C.: Extension principle and fuzzy numbers (II). Fuzzy Mathematics 3, 14–23 (1988) (in Chinese)

[99] Luo, C.: Introduction to Fuzzy Set Theory. Beijing Normal University Press, Beijing (1985) (in Chinese)

[100] Malik, D.S., Mordeson, J.N.: Fuzzy homomrphisms of rings. Fuzzy Sets and Systems 46, 139–146 (1992)

[101] Mamdani, E.H., Assilian, S.: An experiment in linguistic synthesis with a fuzzy logic controller. Int. J. Man Mach. Studies 7, 1–13 (1975)

[102] Mizumoto, M., Zimmermann, J.: Comparison of fuzzy reasoning methods. Fuzzy Sets and Systems 8, 253–283 (1982)

[103] Montero, F.J., Tejada, J.: A necessary and sufficient condition for the existence of Orlovsky's choice set. Fuzzy Sets and Systems 26, 121–125 (1988)

[104] Mordeson, J.N., Malik, D.S.: Fuzzy Commutative Algebra. World Scientific, Singapore (1998)

[105] Mordeson, J.N., Bhutani, K.R., Rosenfeld, A.: Fuzzy Group Theory. Springer, Heidelberg (2005)

[106] Nakamura, K.: Preference relations on a set of fuzzy utilities as a basis for decision making. Fuzzy Sets and Systems 20, 147–162 (1986)

[107] Mukherjee, N.P., Bhattacharya, P.: Fuzzy normal subgroups and fuzzy cosets. Inform. Sci. 34, 225–239 (1984)
[108] Negoita, C.V., Ralescu, D.A.: Applications of Fuzzy Sets to Systems. Birkhauser, Basel (1975)
[109] Orlovsky, S.V.: Decision-making with a fuzzy preference relation. Fuzzy Sets and Systems 1, 155–167 (1978)
[110] Ovchinnikov, S., Roubens, M.: On strict preference relations. Fuzzy Sets and Systems 43, 319–326 (1991)
[111] Pedrycz, W.: Applications of fuzzy relational equations for methods of reasoning in presence of fuzzy data. Fuzzy Sets and Systems 16, 163–175 (1990)
[112] Pedrycz, W.: $s - t$ fuzzy relational equations. Fuzzy Sets and Systems 59, 189–195 (1993)
[113] Pedrycz, W.: Approximate solutions of fuzzy relational equations. Fuzzy Sets and Systems 28, 183–202 (1988)
[114] Peng, Y., Xu, X.: The dual composition of fuzzy relations and its applications to transitivity properties. Fuzzy Systems and Mathematics 19, 54–59 (2005)
[115] Peng, Y., Xu, X., Wu, S.: The relations between transitivity properties in fuzzy preference modelling. Fuzzy Systems and Mathematics 21, 95–99 (2007)
[116] Perfilieva, I., Noskova, L.: System of fuzzy relation equations with inf-→ composition: complete set of solutions. Fuzzy Sets and Systems 159, 2256–2271 (2008)
[117] Pu, P.M., Liu, Y.M.: Fuzzy topology I. J. Math. Anal. Appl. 76, 571–599 (1980)
[118] Rosenfeld, A.: Fuzzy groups. J. Math. Anal. Appl. 35, 512–517 (1971)
[119] Roubens, M., Vincke, P.: Preference Modeling. Springer, Berlin (1985)
[120] Roubens, M.: Some properties of choice functions based on valued binary relations. European Journal of Operations Research 40, 309–321 (1989)
[121] Ralescu, D., Adams, G.: Fuzzy integral. J. Math. Anal. Appl. 75, 562–570 (1986)
[122] Ruan, D., Kerre, E.: Fuzzy implication operators and generalized fuzzy method of cases. Fuzzy Sets and Systems 54, 23–37 (1993)
[123] Ruan, D., Kerre, E.: On the extension of compositional rule of inference. International Journal of Intelligent Systems 8, 807–817 (1993)
[124] Ruan, D., Kerre, E.: On the extended fuzzy chaining sysllogism. In: Proceedings IFSA 1995, vol. 1, pp. 145–148 (1995)
[125] Ramik, J., Rimanek, J.: Inequality relation between fuzzy numbers and its use in fuzzy optimization. Fuzzy Sets and Systems 16, 123–138 (1985)
[126] Saade, J.J., Schwarzlander, H.: Ordering fuzzy sets over the real line: an approach based on decision making under uncertainty. Fuzzy Sets and Systems 50, 237–246 (1992)
[127] Sanchez, E.: Resolution of composite fuzzy relation equations. Information and Control 30, 38–48 (1976)
[128] Sanchez, E.: Solution of fuzzy equations with extended operations. Fuzzy Sets and Systems 12, 237–248 (1984)
[129] Sanchez, E.: Solutions in composite fuzzy relation equations: application to medical diagnosis in Brouwerian logic. In: Gupta, M., Saridis, G., Gains, B. (eds.) Fuzzy Automata and Decision Processes, pp. 221–234. North-Holland, New York (1977)

References

[130] Schweizer, B., Sklar, A.: Probabilistic Metric Spaces. North-Holland, Amsterdam (1983)
[131] Sengupta, K.: Fuzzy preference and Orlovsky choice procedure. Fuzzy Sets and Systems 93, 231–234 (1998)
[132] Sherwood, H.: Products of fuzzy subgroups. Fuzzy Sets and Systems 11, 79–89 (1983)
[133] Shi, Y., Ruan, D., Kerre, E.: On the characterizations of fuzzy implications satisfying $I(x,y) = I(x, I(x,y))$. Information Sciences 177, 2954–2970 (2007)
[134] Shi, Y., Van Gasse, B., Ruan, D., Kerre, E.: On the first place antitonicity in QL-implications. Fuzzy Sets and Systems 159, 2988–3013 (2008)
[135] Smets, P., Magrez, P.: Implication in fuzzy logic. Int. J. Approximate Reasoning 1, 327–347 (1987)
[136] Suarez, G., Gilavarez, P.: Two families of fuzzy integrals. Fuzzy Sets and Systems 18, 67–82 (1986)
[137] Sugeno, M.: Theory of fuzzy integrals and its applications, Ph.D. Thesis, Tokyo Institute of Technology (1974)
[138] Sugeno, M. (ed.): Industrial Applications of Fuzzy Control. North-Holland, New York (1985)
[139] Sugeno, M., Nishida, M.: Fuzzy control of model car. Fuzzy Sets and Systems 16, 103–113 (1985)
[140] Tang, F.: Fuzzy bilinear equations. Fuzzy Sets and Systems 28, 217–226 (1988)
[141] Terano, T., Asai, K., Sugeno, M. (eds.): Applied Fuzzy Systems. Academic Press, London (1994)
[142] Tong, R.M., Bonissone, P.P.: A linguistic approach between fuzzy numbers and its use in fuzzy sets. IEEE Trans. Systems Man Cybernet. 10, 716–723 (1980)
[143] Van Leekwijck, W., Kerre, E.: Defuzzification: criteria and classification. Fuzzy Sets and Systems 108, 159–178 (1999)
[144] Wagenknecht, M., Hartmann, K.: On the existence of minimal solutions for fuzzy equations with tolerances. Fuzzy Sets and Systems 34, 237–244 (1990)
[145] Wang, X., Kerre, E., Ruan, D.: Consistency of judgement matrix and fuzzy weights in fuzzy analytic hierarchy process. Int. J. uncertainty, Fuzziness and Knowledge-based Syst. 3, 35–46 (1995)
[146] Wang, X., De Baets, B., Kerre, E.: A comparative study of similarity measures. Fuzzy Sets and Systems 73, 259–268 (1995)
[147] Wang, X., Ruan, D.: On the transitivity of fuzzy preference relations in ranking fuzzy numbers. In: Ruan, D. (ed.) Fuzzy Set Theory and Advanced Mathematical Applications, pp. 155–173. Kluwer Academic Publishers, Dordrecht (1995)
[148] Wang, X., Kerre, E.: On the classification and the dependencies of the ordering methods. In: Ruan, D. (ed.) Fuzzy Logic Foundations and Industrial Applications, pp. 73–88. Kluwer Academic Publishers, Dordrecht (1996)
[149] Wang, X., Kerre, E.: Reasonable properties for the ordering of fuzzy quantities (I). Fuzzy Sets and Systems 118, 375–385 (2001)
[150] Wang, X., Kerre, E.: Reasonable properties for the ordering of fuzzy quantities (II). Fuzzy Sets and Systems 118, 387–405 (2001)
[151] Wang, X.: A comparative study of the ranking methods for fuzzy quantities, Ph. D Thesis, Ghent University (1997)

[152] Wang, L.X.: A Course in Fuzzy Systems and Control. Prentice-Hall, NJ (1996)
[153] Wang, Z.: The autocontinuity of set function and the fuzzy integral. J. Math. Anal. Appl. 99, 195–218 (1984)
[154] Wang, Z.: Notes on the convergence theorem of sequences of fuzzy integrals in measure. Fuzzy Mathematics 4, 7–10 (1984) (in Chinese)
[155] Wang, Z.: Asymptotic structural characteristics of fuzzy measure and their application. Fuzzy Sets and Systems 16, 277–290 (1985)
[156] Wang, Z., Klir, G.: Fuzzy Measure Theory. Plenum Press, New York (1992)
[157] Warren, R.H.: Neighborhoods, bases and continuity in fuzzy topological spaces. Rocky Mt. J. Math. 8, 459–470 (1978)
[158] Watson, S.R., Weiss, J.J., Donnel, M.L.: Fuzzy decision analysis. IEEE Trans. Systems Man Cybernet 9, 1–9 (1979)
[159] Wu, W.M.: Normal fuzzy subgroups. Fuzzy Mathematics 1, 21–30 (1981) (in Chinese)
[160] Wu, C., Ma, M.: Structure Theory of Fuzzy Analsis. International Academic Press, London (1990)
[161] Wu, C., Wang, S., Song, S.: Generalized triangle norms and generlized fuzzy integrals. In: Proceedings of Sino-Japan Symposium on Fuzzy Sets and Systems. International academic Press, London (1990)
[162] Yager, R., Rybalov, A.: Uninorm aggregation opertors. Fuzzy Sets and Systems 80, 111–120 (1996)
[163] Yager, R.R.: On choosing between fuzzy subsets. Kybernetes 9, 151–154 (1980)
[164] Yager, R.R.: A procedure for ordering fuzzy sets of the unit interval. Information Sciences 24, 143–161 (1981)
[165] Yager, R.R.: Ranking fuzzy subsets over the unit interval. In: Proc. 1978 CDC, pp. 1435–1437 (1978)
[166] Yagishita, O., Itoh, O., Sugeno, M.: Application of fuzzy reasoning to the water purification process. In: Sugeno, M. (ed.) Industrial Applications of Fuzzy Control, pp. 19–40. North-Holland, Amsterdam (1985)
[167] Yamakawa, T.: Stabilization of an inverted pendulum by a high-speed fuzzy logic controller hardware systems. Fuzzy Sets and Systems 32, 161–180 (1989)
[168] Yasunobu, S., Miyamoto, S.: Automatic train operation by predictive fuzzy control. In: Sugeno, M. (ed.) Industrial Application of Fuzzy Control, pp. 1–18. North-Holland, Amsterdam (1985)
[169] Yu, Y.: A theory of isomorphisms of fuzzy groups. Fuzzy Syst. and Math. 2, 57–68 (1988) (in Chinese)
[170] Yuan, Y.: Criteria for evaluating fuzzy ranking methods. Fuzzy Sets and Systems 43, 139–157 (1991)
[171] Zadeh, L.A.: Fuzzy sets. Information and Control 8, 338–353 (1965)
[172] Zadeh, L.A.: The concept of a linguistic variable and its application to approximate reasoning-I. Information Sciences 8, 199–251 (1975)
[173] Zadeh, L.A.: The concept of a linguistic variable and its application to approximate reasoning-II. Information Sciences 8, 301–357 (1975)
[174] Zadeh, L.A.: The concept of a linguistic variable and its application to approximate reasoning-III. Information Sciences 9, 43–80 (1975)
[175] Zadeh, L.A.: Similarity relations and fuzzy orderings. Information Sciences 3, 177–200 (1971)

References

[176] Zadeh, L.A.: Fuzzy sets as a basis for a theory of possibility. Fuzzy Sets and Systems 1, 3–28 (1978)

[177] Zenner, R.B., Kerre, E.E., De Caluwe, R.: Practical determination of document description relations in document retrieval systems. In: Carlsson, C. (ed.) Proceedings of the Workshop on the Membership Function, EIASM, Brussels, pp. 127–138 (1984)

[178] Zhang, G.: Fuzzy Measure. Guizhou Science and Technology Press (1992) (in Chinese)

[179] Zhang, W.: Fundamentals of fuzzy mathematics. Xian Jiaotong University Press, Xian (1984) (in Chinese)

[180] Zhang, W., Liang, G.: Fuzzy Control and Fuzzy Systems. Xian Jiaotong University Publishing House, Xian (1986) (in Chinese)

[181] Zhao, R.: (N)-fuzzy integral. Journal of Mathematical Research and Exposition 2, 55–72 (1981) (in Chinese)

Index

L-fuzzy set 47
P-closure 78
R-afterset 66
R-foreset 66
R-implication 34
S-completeness 85
S-implication 33
S-membership function 128
T-S-Ferrers relation 85
T-S-semitransitivity 85
T-antisymmetry 85
T-composition 72
α-cut 39
α-cut relation 67
α-generated fuzzy topological space 177
λ-fuzzy measure 155
n-ary extension principle 121
(fuzzy) negation 26

acyclicity 91
afterset 3
algebraic operation 8
a nest of closed intervals 129
a nest of intervals 125
a nest of sets 44
antisymmetry 10, 83
atomic proposition 191
automorphism 27

base 177
bijection 8
Boolean algebra 14
Boolean lattice 14
bounded lattice 13

cartesian product of fuzzy sets 119
characteristic function 2
closed (fuzzy) set 176
closure 176
complemented lattice 14
complement of fuzzy set 23
complete lattice 15
completely distributive lattice 15
completeness 83
complete normality 185
composition of fuzzy relations 69
composition of mappings 7
composition of relations 4
compound proposition 191
concentration 190
condition ($\mathcal{L}$) 52
conjunction 28
consistence 91
convex fuzzy quantity 124
convex fuzzy set 123
coset 168
crisp set 22

Decomposition Theorem I 42
Decomposition Theorem II 42
Decomposition Theorem III 43
defuzzification 199
degree of membership 22
De Morgan triple 31
dilatation 190
disjunction 29
distributive lattice 13
dual intuitionistic negation 26

epimorphism   9
equivalence class   6
equivalence relation   6
extension principle   115

Ferrers relation   83
foreset   3
fuzzification   198
fuzzy T-transitivity   75
fuzzy (binary) relation   65
fuzzy clustering analysis   99
fuzzy control   201
fuzzy equivalence   38
fuzzy equivalence relation   75
fuzzy ideal   173
fuzzy IF-THEN rule   191
fuzzy implication   32
fuzzy integral   157
fuzzy measurable space   153
fuzzy measure   153
fuzzy measure space   153
fuzzy number   126
fuzzy point   176
fuzzy proposition   191
fuzzy quantity   122
fuzzy relation equation   94
fuzzy set   22
fuzzy Sierpinski space   185
fuzzy singleton   176
fuzzy subgroup   163
fuzzy subring   171
fuzzy tolerance relation   80
fuzzy topological space   176
fuzzy topology   175

Gaussian fuzzification   199
Generalized hypothetical syllogism   194
Generalized modus ponens   194
Generalized modus tollens   194
Goguen equivalence   38

hedge   190
height   25
homomorphism   8
Hutton-normality   184
hypothetical syllogism   193

image   7
included fuzzy singleton fuzzy topology   185

induced fuzzy topology   185
infimum   11
injection   8
inner product   58
interior   176
intersection of fuzzy sets   23
intuitionistic negation   26
inverse of relation   4
involution   26
irreflexivity   85
isomorphism   9

kernel of fuzzy set   25
Kerre-normality   185
Kerre neighborhood   176
Kleene-Dienes implication   33

largest fuzzy topology   185
lattice   12
lattice nearness measure   58
left coset   168
linguistic variable   189
Lowen or stratified fuzzy topological space   177
lower bound   11
Ludescher neighborhood   176
Lukasiewicz equivalence   38
Lukasiewicz implication   33

mapping   7
Mashhour neighborhood   176
maximal solution   94
modus ponens   193
modus tollens   193
Morgan algebra   14
Morsi fuzzy topological space   177

nearness measure   56
negative S-transitivity   85
negative transitivity   83
non-negative fuzzy quantity   122
normal fuzzy subgroups   167
null-additivity   187

operation   8
outer product   58

partial order   10
plinth   25
poset   10

Index

positive fuzzy quantity   122
pseudo-complement   46
Pu neighborhood   176

quasi-coincidence   176
quotient group   169
quotient ring   175
quotient set   6

recurring   195
reflexive (fuzzy) relation   74
reflexivity   6, 10
Reichenbach implication   33
relation   3
Representation Theorem   44
residual implication   34
right coset   168

semitransitivity   83
simple function   159
smallest fuzzy topology   185
soft algebra   14
strict De Morgan triple   31
strict negation   26
strong α-cut   39
strong α-cut relation   67
strong De Morgan triple   31
strong negation   26
subbase   177
subcomposition of fuzzy relations   72
subcomposition of relations   5

supercomposition of fuzzy relations   72
supercomposition of relations   5
superior soft algebra   16
support   25
supremum   10
surjection   8
surjective fuzzy topological space   177
symmetric (fuzzy) relation   74
symmetry   6

topologically generated fuzzy topological space   177
transitive (fuzzy) relation   75
transitivity   6, 10
trapezoidal fuzzy number   128
triangular composition of relations   5
triangular conorm (t−conorm)   29
triangular fuzzification   199
triangular fuzzy number   128
triangular norm (t-norm)   28
type-2 fuzzy set   49

uninorm   32, 62
union of fuzzy sets   23
upper bound   10

Warren neighborhood   176
weak normality   185
weak transitivity   91

Breinigsville, PA USA
02 November 2009
226921BV00005B/112/P